MATH

Grade 8

Thomas J. Richards
Mathematics Teacher
Lamar Junior-Senior High School
Lamar, Missouri

This book is dedicated to our children — Alyx, Nathan, Fred S., Dawn, Molly, Ellen, Rashaun, Brianna, Michele, Bradley, BriAnne, Kristie, Caroline, Dominic, Corey, Lindsey, Spencer, Morgan, Brooke, Cody, Sydney — and to all children who deserve a good education and who love to learn.

McGraw-Hill Consumer Products

McGraw-Hill
Consumer Products

A Division of The McGraw-Hill Companies

Copyright © 1997 McGraw-Hill Consumer Products.
Published by McGraw-Hill Learning Materials, an imprint of
McGraw-Hill Consumer Products.

Send all inquiries to:
McGraw-Hill Consumer Products
250 Old Wilson Bridge Road
Worthington OH 43085

ISBN 1-57768-118-5

3 4 5 6 7 8 9 10 POH 03 02 01 00 99

Table of Contents

The SPECTRUM

MATHEMATICS Series

Using This Book

SPECTRUM MATHEMATICS is a non-graded, consumable series for students who need special help with the basic skills of computation and problem solving. This successful series emphasizes skill development and practice, without complex terminology or abstract symbolism. Because of the nature of the content and the students for whom the series is intended, readability has been carefully controlled to comply with the mathematics level of each book.

Features:

- A **Pre-Test** at the beginning of each chapter helps determine a student's understanding of the chapter content. The Pre-Test enables students and teachers to identify specific skills that need attention.
- **Developmental exercises** are provided at the top of the page when new skills are introduced. These exercises involve students in learning and serve as an aid for individualized instruction or independent study.
- **Abundant opportunities for practice** follow the developmental exercises.
- **Problem-solving pages** enable students to apply skills to realistic problems they will meet in everyday life.

- A **Test** at the end of each chapter gives students and teachers an opportunity to check understanding. A **Mid-Book Test**, covering Chapters 1–5, and a **Final Test**, covering all chapters, provide for further checks of understanding.
- A **Record of Test Scores** is provided on page xvi of this book so students can chart their progress as they complete each chapter test.
- **Answers** to all problems and test items are included at the back of the book.

This is the third edition of *SPECTRUM MATHEMATICS*. The basic books have remained the same. Some new, useful features have been added.

New Features:

- **Scope and Sequence Charts** for the entire *Spectrum Mathematics* series are included on pages iv–v.
- **Problem-Solving Strategy Lessons** are included on pages vii–xiv. These pages may be used at any time. The purpose is to provide students with various approaches to problem-solving.
- An **Assignment Record Sheet** is provided on page xv.

Problem-Solving Strategies

NAME _____

To solve problems, you must be able to **choose the correct operation**.

It is also helpful to have a plan to solve a problem. **Plan**

Janita earned $18,150. She paid $3,205 in taxes. She saved $3,785. She spent the rest of the money. How much did she spend?

Strategy: Add to find how much she used for taxes and saving. Subtract to find how much she had left to spend.

Janita spent $ _____.

Plan	Read the problem.

Read the problem.
Identify the question.
Identify the information you
    need to solve the problem.
Choose the correct operation.
Check that the answer is reasonable.

You can **estimate** to check for a reasonable answer.
Earned − (Taxes + Savings) = Spent
$18,000 − ($3,000 + 4,000) = $11,000
Your answer should be about $11,000.

Solve each problem.

1. Each box holds 23 parts. How many boxes are needed to hold 3,105 parts?

Each box holds _____ parts.

There are _____ parts.

To solve the problem, will you add, subtract, multiply, or divide?

_____ to solve the problem.

_____ boxes are needed to hold 3,105 parts.

Is your answer reasonable? _____

2. You buy items that cost $41.43, $25.49, and $71.94. The tax is $9.03. What is the total of the items you bought including tax?

To solve the problem, will you add, subtract, multiply, or divide?

_____ to solve the problem.

The total cost is $ _____.

3. Jack is 14 years old. Five years ago, his uncle was twice as old as Jack was then. How old is Jack's uncle now?

Jack's uncle is _____ years old now.

1.

2.

3.

Perfect score: 8   My score: _____

Solve each problem.

1. There were 423 men, 328 women, and 148 children at the theater. Only 523 people paid to get in the theater. The rest of the people were using tickets they had won. How many people won tickets to get in the theater?

Will you add, subtract, multiply, or divide to find how many people are at the theater?

_____ to find how many people are at the theater.

Now, to solve the problem, will you add, subtract, multiply, or divide?

_____ to solve the problem.

_____ people won tickets to get in the theater.

Is your answer reasonable? _____

2. A bus travels at an average of 45 miles per hour. At that rate, how long will it take a bus to travel 585 miles, if the bus makes two stops, each of which takes 30 minutes?

It will take _____ hours.

3. How long would it take the bus in 2 if it made 6 stops, each for 20 minutes?

It would take _____ hours.

4. Marion worked 40 hours last week. He earned $480. At that rate of pay, how much would he be paid if he had only worked 36 hours?

For 36 hours, he would have earned $ _____.

5. Machine A cost $5,500 and makes 425 parts each hour. Machine B cost $8,200 and makes 600 parts each hour. How much more did Machine B cost than Machine A?

Machine B cost $ _____ more than Machine A.

6. Each of the machines in 5 operates for 8 hours each day. At the end of one day, how many parts in all will both machines make?

Together the machines will make _____ parts in one day.

1.

2–3.

4.

5–6.

Perfect score: 9   My score: _____

Problem-Solving Strategies

Sometimes you can **make a list** to solve problems.

Committee Members
Chris
Arleta
Benjamin
Merilynne
Heather

You are to choose 2 people to attend a meeting.

How many different combinations of people are there?
Make a list to solve the problem.

Chris, Arleta	Arleta, Benjamin	Benjamin, Heather
Chris, Benjamin	Arleta, Merilynne	Merilynne, Heather
Chris, Merilynne	Arleta, Heather	
Chris, Heather	Benjamin, Merilynne	

There are _____ different combinations.

Make a list to solve each problem.   *Lists*

1. There are 8 people at a meeting. Each person is to shake hands with every other person. Make a list to show all the possible combinations of 2 people shaking hands. To make it easier, use letters instead of names. Use A, B, C, D, E, F, G, and H.

There are _____ possible hand-shaking combinations of 2 people.

2. You have 1 penny, 1 nickel, 1 dime, and 1 quarter. How many different amounts of money can you make? You can use from 1 to 4 coins for each amount. Show your list at the right.

I can make _____ different amounts.

3. You have these cards: [V] [W] [X] [Y] [Z]
You are to draw 3 cards. How many different combinations of 3 cards can you make? Show your list at the right.

I can make _____ different combinations.

4. Your company needs to hire new delivery drivers. Make a list of qualifications you would expect the drivers to have.

_____

_____

Compare your list with the list of other classmates. How did making the list make it possible to solve the problem of hiring new delivery drivers?

Perfect score: 4   My score: _____

Problem-Solving Strategies

Sometimes you can **make a table** to solve problems.

In 1 hour, Machine A makes 18 parts. How many parts can the machine make in 5 hours? How long will it take to make 54 parts? 72 parts?

Making a table is useful when you are asked more than one question, based on given information.

Number of hours	1	2	3	4	5
Number of parts	18	36	54	72	

1×18 2×18 3×18 4×18 5×18

Complete the table for 5 hours. Then use the table to answer each question.

It will take _____ hours to make 54 parts.

It will take _____ hours to make 72 parts.

Make a table for each problem. Then use the table to answer each question.

1. Barb made 100 units in 4 days. She made the same number of units each day. How many units did she make in 5 days? How long would it take her to make 200 units?

1.

Number of days	1	2	3	4		
Number of units	25	50				

$100 \div 4$ 2×25 3×25

She made _____ units in 5 days.

It would take _____ days to make 200 units.

2. Ten people are on a bus. At each of the next 6 stops, 4 more people get on. No one gets off. How many people are on the bus after the fifth stop? after the sixth stop? At which stop are 22 people on the bus?

2.

Number of stops	1	2	3			
Number of people on	14	18				

$10 + 4$ $14 + 4$ $18 + 4$

_____ people are on after the fifth stop.

_____ people are on after the sixth stop.

Twenty-two people are on after the _____ stop.

Perfect score: 18   My score: _____

Problem-Solving Strategies

Sometimes you can **make or use a drawing** to solve problems.

Lila has a garden that is 12 feet wide and 5 feet long. She plants 12-feet rows across the garden. She starts and finishes the rows on each edge of the garden. If the rows are 1 foot apart, how many rows will there be? Make a drawing, then answer the question.

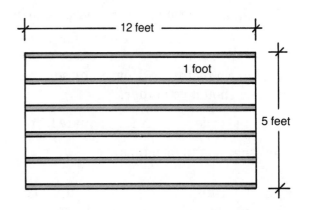

There will be _____ rows.

Make or use a drawing to solve each problem.

1. Mark put posts along the fence line. Because the ground was rocky, he couldn't space the posts evenly. Post A is put at the beginning of the fence line. Post B is 8 feet east of A. Post C is 9 feet east of B. Post D is 10 feet east of C. Post E is 6 feet east of D. How far is post E from A?

Post E is _____ feet from post A.

1.

2. June is cutting parts for a machine. She has the drawing at the right showing the sizes of pieces needed for the machine. Use the drawing. What is the length and width of the piece of metal that the parts will be cut from?

The piece of metal is _____ feet long and _____ feet wide.

2.

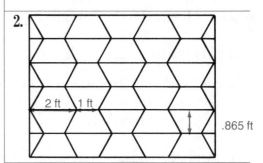

3. Five people are in line. Abel is right behind Bea. Carol is in front of Dana, but not right in front of Dana. Dana is behind Bea. Ed is last in line. Where is Dana in the line?

Dana is _____ in line.

3.

4. Melanie put a fence around her rectangular garden. The garden is 10 feet by 12 feet. She placed a post at each corner. The rest of the posts were 2 feet apart. How many posts did she use?

She used _____ posts.

4.

Perfect score: 5   My score: _____

Problem-Solving Strategies

You can **use an equation or a formula** to help solve problems.

You are ordering new carpet for your office. The office is 24 feet long and 18 feet wide and is shaped like a rectangle. How many square feet of carpeting should you order?

Use the formula for the area of a rectangle.

Area = length × width

You should order _____ square feet.

Some other useful formulas are:

A = l × w
A = 24 × 18
A = _____

distance = rate × time   area of a triangle = ½ × base × height

volume of a rectangular prism = length × width × height

Use a formula to solve each problem.

1. You are fertilizing a flower garden shaped like a triangle. The base of the triangle is 8 feet and the height is 4 feet. What is the area of the flower garden?

   The area is _____ square feet.

   1.

2. Julia is planning a trip. She is going to a town 336 miles away. She plans to make 1 stop for gas and food that will take her about ½ hour. She thinks she can average 48 miles per hour. How long will it take to complete her trip?

   It will take her _____ hours.

   2.

3. Alicia is going to carpet her rectangular living room. The room is 12 feet long and 15 feet wide. How many square feet of carpeting will Alicia need?

   She will need _____ square feet of carpet.

   3–4.

4. The carpet Alicia (in 3) picked out sells for $20 per square yard. How many square yards of carpet will she need? How much will the carpet cost?

   She will need _____ square yards.

   It will cost $ _____.

5. A company has a large storage area. It is a rectangular prism that is 150 feet long, 60 feet wide, and 10 feet tall. How many cubic feet of space is in the storage area?

   The storage area has _____ cubic feet.

   5.

Perfect score: 6   My score: _____

Problem-Solving Strategies

Many times there is more than one way to solve a problem. You can **choose the strategy** that you like to use best.

Remember to follow the problem-solving plan.

Train A leaves Ourtown and travels at 90 miles per hour. Train B leaves Yourtown and travels 120 miles per hour. The trains are traveling toward each other and pass each other in 2 hours. How far apart are the towns of Ourtown and Yourtown?

You could make a list or a table.

Train	1 Hour	2 Hours
A	90	180
B	120	240

Plan

Read the problem.
Identify the question.
Identify the information you
   need to solve the problem.
Decide what strategy to use.
Use that strategy to find the
   answer.
Does your answer make sense?

You could make a drawing.

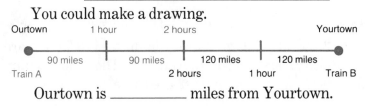

Ourtown is _____ miles from Yourtown.

Choose a strategy. Then solve each problem.

1. Alan ran 8 miles Monday and 6 miles Tuesday. Marlene ran half as far on Monday as Alan ran on Monday. On Tuesday Marlene ran twice as far as Alan. How far did each person run in all on those two days?

Alan ran _____ miles in all on the 2 days.

Marlene ran _____ miles in all on the 2 days.

2. The Browns started driving toward the Edwards' house. The Browns drove for 3 hours at an average speed of 50 miles per hour. Then they ran into heavy city traffic. Mrs. Brown said that it would take them 2 more hours if they averaged 33 miles per hour. How far were the Browns from the Edwards' house when they started driving?

They were _____ miles from the Edwards' house when they started.

3. A frog starts at the bottom of a steep hill. It jumps 3 feet forward on each jump and then slips back 1 foot before it can jump again. It is 18½ feet from the bottom of the hill to the top. How many jumps must the frog make to get to the top of the hill?

The frog must make _____ jumps.

1.

2.

3.

Perfect score: 4   My score: _____

Problem-Solving Strategies

Decide what strategy or operation to use to solve each problem. Then solve.

1. Thirty people are on a bus. Five people get off and two people get on at each stop. How many people are on the bus after the fifth stop?

_____ people are still on the bus.

1.

2. Your group is planning a community meeting. Each person in the group is to be on two committees. The committees are: Refreshments, Decorations, Cleanup, and Invitations. How many different combinations of 2 committees are there?

There are _____ different combinations.

3. In 2, what if two more committees (Child Care and Publicity) are added to the list. Then how many combinations of two committees are there?

There are _____ different combinations.

2–3.

4. Zeta Company made $158,000 last year. Taxes were $68,400. This year they made 3 times as much as last year, but their taxes were 5 times higher. How much money did Zeta Company have after paying taxes this year?

Zeta Company had $ _____ after taxes this year.

4.

5. A train travels 140 miles in 2 hours. At that rate, how far will it travel in 1 hour? in 6 hours?

In 1 hour, it will travel _____ miles.

In 6 hours, it will travel _____ miles.

5.

6. During a meeting the following exchanges took place. Arnie gave Glenn $44. Xenia gave Ursula $32. Alan gave Xenia $18. Alan also gave Glenn and Ursula each $20. Glenn gave Arnie $30 and Alan $10. Find how much each person started with and ended with.

6.

Name	Started With	Ended With
Arnie		
Glenn		
Xenia		
Ursula		
Alan		

Perfect score: 16   My score: _____

NAME _____

Assignment Record Sheet

Pages Assigned	Date	Score	Pages Assigned	Date	Score	Pages Assigned	Date	Score

SPECTRUM MATHEMATICS

Record of Test Scores

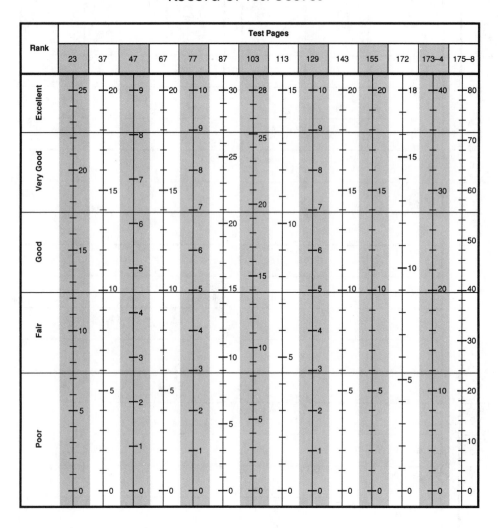

To record the score you receive on a TEST:

  (1) Find the vertical scale below the page number of that TEST,
  (2) on that vertical scale, draw a ● at the mark which represents
      your score.

For example, if your score for the TEST on page 23 is "My score: 15," draw a ● at the 15-mark on the first vertical scale. A score of 15 would show that your rank is "Good." You can check your progress from one test to the next by connecting the dots with a line segment.

PRE-TEST—Addition, Subtraction, Multiplication, and Division

Add or subtract. Write each answer in simplest form.

	a	b	c	d	e
1.	2 3 1 2 +5 6 7 8	1 2.3 +4.8	1 5 7 8 2 3 2 1 4 2 5 +2 1 4 2 8	5 6.7 3 4.1 8 +3.1 4	8 9.5 7 8 2 1.3 4 5 +3 4.5 7 9
2.	6 7 5 2 −9 7 4	7.2 −3.4	2 8 7 5 2 8 −3 1 6 7 8	1 6.0 1 −8.6 4	2 5 7.3 1 4 −1 5 1.2 7 6

3.

a. $\dfrac{7}{10}$
   $+\dfrac{4}{10}$

b. $\dfrac{3}{4}$
   $-\dfrac{1}{3}$

c. $3\dfrac{2}{3}$
   $+1\dfrac{1}{4}$

d. $5\dfrac{1}{5}$
   $-2\dfrac{5}{6}$

e. $4\dfrac{1}{2}$
   $3\dfrac{1}{3}$
   $+4\dfrac{3}{8}$

Multiply or divide. Write each answer in simplest form.

	a	b	c	d
4.	3 5 1 ×2 8	8 3 4 ×7 8 2	1 5.2 4 ×3.6	5.6 7 8 ×1 2.3

5.

a. $\dfrac{1}{2}\times\dfrac{2}{3}$

b. $\dfrac{3}{8}\div\dfrac{2}{3}$

c. $1\dfrac{2}{3}\times\dfrac{5}{6}$

d. $2\dfrac{1}{2}\div3\dfrac{1}{8}$

6. $82\overline{)9\ 8\ 9}$ $123\overline{)1\ 4\ 5\ 6}$ $.096\overline{)4\ 7.5\ 2}$ $2.42\overline{)7.8\ 6\ 5\ 0}$

Perfect score: 27   My score: _____

1

PRE-TEST Problem Solving

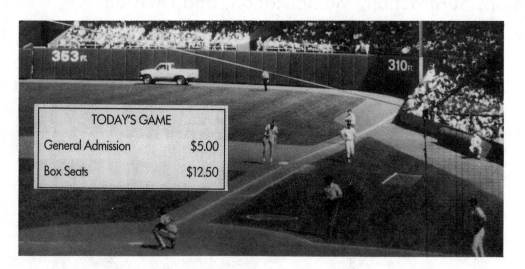

Solve each problem.

1. There were 3,428 box-seat tickets sold for today's ball game. How much money was collected for the box-seat tickets sold?

$_____ was collected.

2. There were 9,347 general-admission tickets sold. How much money was collected for the general-admission tickets sold?

$_____ was collected.

3. What was the total number of box-seat and general-admission tickets sold for today's ball game?

The total number sold was _____.

4. For today's ball game, how many more general-admission tickets were sold than box-seat tickets?

_____ more were sold.

5. At one of the ticket booths, a total of $1,690 was collected for general-admission tickets. How many general-admission tickets were sold at that booth?

_____ were sold.

1.

2.

3.

4.

5.

Perfect score: 5   My score: _____

Lesson 1 Addition

Add the ones. Rename 27 as "2 tens and 7 ones."

Continue adding from right to left.

$$
\begin{array}{r}
2\ 1\ 2\ 1\ \overset{2}{0}\ 4 \\
3\ 2\ 3\ 6\ 1\ 6 \\
1\ 3\ 2\ 4\ 0\ 8 \\
+2\ 4\ 1\ 7\ 5\ 9 \\
\hline
7
\end{array}
$$

$\longrightarrow$

$$
\begin{array}{r}
\overset{1}{2}\ 1\ \overset{1}{2}\ 1\ \overset{2}{0}\ 4 \\
3\ 2\ 3\ 6\ 1\ 6 \\
1\ 3\ 2\ 4\ 0\ 8 \\
+2\ 4\ 1\ 7\ 5\ 9 \\
\hline
9\ 0\ 9\ 8\ 8\ 7
\end{array}
$$

Add.

	a	*b*	*c*	*d*	*e*
1.	23 +14	224 +73	312 +5324	43214 +4325	41321 +612314
2.	16 47 +13	217 316 +142	3317 2154 +1212	21016 14527 +51202	260316 217327 +411342
3.	31 70 14 +52	273 162 253 +210	1131 2262 3473 +1051	41370 2151 33225 +11118	121065 302432 304144 +41213
4.	36 75 84 31 +17	633 710 821 502 +221	1123 2651 1762 2873 +1411	11616 12573 21412 40331 +13214	232362 351171 64221 71141 +182314
5.	34 76 58 67 +73	542 624 852 715 +316	7067 8458 5312 2521 +1214	31145 14214 3142 76125 +3214	212304 321456 214672 523214 +314235

Perfect score: 25   My score: _____

Lesson 2   Subtraction

To subtract ones, rename
3 tens and 2 ones as
"2 tens and 12 ones."

Continue subtracting from
right to left, renaming
as necessary.

$$9\ 6\ 5\ 4\ 3\overset{2}{\cancel{3}}\ \overset{12}{\cancel{2}}$$
$$-1\ 2\ 1\ 7\ 1\ 5$$
$$7$$

$$\longrightarrow$$

$$9\ 6\ \overset{4}{\cancel{5}}\ \overset{14}{\cancel{4}}\ \overset{2}{\cancel{3}}\ \overset{12}{\cancel{2}}$$
$$1\ 2\ 1\ 7\ 1\ 5$$
$$8\ 4\ 3\ 7\ 1\ 7$$

Subtract.

	a	b	c	d	e
1.	37 −6	327 −16	4325 −214	17625 −3214	321459 −20123
2.	59 −14	847 −231	6875 −1534	87654 −14123	582785 −131524
3.	70 −49	968 −159	8752 −4127	78547 −31218	495627 −314518
4.	97 −78	523 −172	5963 −2172	25753 −14182	457245 −112158
5.	83 −45	675 −289	5028 −4917	86743 −21892	675247 −321482
6.	45 −27	607 −299	8207 −3149	74003 −21456	900435 −417624
7.	81 −27	700 −287	6732 −865	67524 −29689	351257 −165268

Perfect score: 35   My score: _____

4

Lesson 3   Multiplication

$$
\begin{array}{r}
4873 \\
\times 296 \\
\hline
29238 \\
438570 \\
974600 \\
\hline
1442408
\end{array}
$$

—————— 6 × 4873
—————— 90 × 4873
—————— 200 × 4873

Add.

Multiply.

	a	*b*	*c*	*d*	*e*
1.	63 ×4	432 ×2	679 ×7	2312 ×3	7598 ×8
2.	68 ×20	700 ×34	212 ×43	1720 ×64	2806 ×97
3.	341 ×200	213 ×320	403 ×212	1414 ×312	5875 ×678
4.	700 ×426	646 ×925	925 ×436	9251 ×809	7487 ×869

Perfect score: 20   My score: _____

Lesson 4   Division

Think

$$312 \times 1000 = 312000$$
$$312 \times 100 = 31200$$

Quotient is between 100 and 1000. So its first digit will be in *hundreds* place.

$$312\overline{)66831}$$

$3\overline{)6}$ is 2.

```
       2
312 )66831
     62400
      4431
```

$3\overline{)4}$ is about 1.

```
      21
312 )66831
     62400
      4431
      3120
      1311
```

$3\overline{)13}$ is about 4.

```
     214 r63
312 )66831
     62400
      4431
      3120
      1311
      1248  remainder
        63
```

Divide.

	a	b	c	d	e
1.	$9\overline{)687}$	$42\overline{)367}$	$58\overline{)696}$	$123\overline{)975}$	$421\overline{)2354}$
2.	$7\overline{)1425}$	$57\overline{)1457}$	$69\overline{)8345}$	$521\overline{)7295}$	$624\overline{)12354}$
3.	$6\overline{)37524}$	$83\overline{)12576}$	$37\overline{)84576}$	$784\overline{)79984}$	$379\overline{)97542}$

Perfect score: 15   My score: _____

Lesson 5   Problem Solving

Solve each problem.

1. Luke delivered 225 circulars on Taft Street, 134 on 87th Street, 218 on Roosevelt Street, and 229 on 89th Street. How many circulars did he deliver?

He delivered _____ circulars.

2. Babe Ruth hit 714 home runs during his career. Lou Gehrig hit 493. How many more home runs did Ruth hit than Gehrig?

Ruth hit _____ more home runs.

3. If Mrs. Jones drives 350 kilometers each day, how many kilometers will she drive in 4 days?

She will drive _____ kilometers.

4. A bricklayer laid 656 bricks in 8 hours. Suppose the same number of bricks were laid each hour. How many bricks were laid each hour?

_____ bricks were laid each hour.

5. The load limit for a small bridge is 6,000 pounds. Mr. Sims' car weighs 4,175 pounds. How much less than the load limit does the car weigh?

The car weighs _____ pounds less.

6. One week the bakery used 144 sacks of flour. Suppose each sack of flour weighed 125 pounds. What was the total weight of the flour used?

_____ pounds of flour were used.

7. Seventeen thousand three hundred seventy items are to be packed into boxes of 24 items each. How many boxes will be filled? How many items will be left over?

_____ boxes will be filled.

_____ items will be left over.

1.	2.
3.	4.
5.	6.
7.	

Perfect score: 8   My score: _____

Problem Solving

Solve each problem.

1. The local Plumbers Union has 456 members. The local Carpenters Union has 875. How many more members does the Carpenters Union have than the Plumbers Union?

There are _____ more members in the Carpenters Union.

2. Seven cars can be loaded on a transport truck. Each car weighs 4,125 pounds. What is the total weight of the cars that can be loaded on the truck?

The total weight is _____ pounds.

3. Mr. Cosgrove drove his car 10,462 miles last year and 11,125 miles this year. How many miles did he drive the car during these 2 years?

He drove the car _____ miles.

4. There were 1,172 women at a banquet. They were seated 8 to a table. How many tables were filled? How many were at the partially-filled table?

_____ tables were filled.

_____ women were at the partially filled table.

5. There are 2,125 employees at the McKee Plant. Each works 35 hours a week. What is the total number of hours worked by these employees in 1 week?

The total number of hours is _____.

6. The population of Tomstown is 34,496 and the population of Jacksburg is 28,574. How much greater is the population of Tomstown than the population of Jacksburg?

The population is _____ greater.

7. A satellite is orbiting the earth at a speed of 17,640 miles an hour. At this rate how many miles will the satellite travel in 1 minute?

It will travel _____ miles.

1.	2.
3.	**4.**
5.	**6.**
7.	

Perfect score: 8   My score: _____

8

Lesson 6   Addition and Subtraction

When adding or subtracting decimals, write the decimals so the decimal points line up. Then add or subtract as with whole numbers.

```
  26.94        26.940  ← Write these
  45.836       45.836    0's if they
+ 32         + 32.000    help you.
            ----------
            104.776
```

```
   62.5         62.500  ← Write these
 -43.345      -43.345      0's if they
            ----------     help you.
              19.155
```

The decimal point in the answer is directly below the other decimal points.

Add or subtract.

	a	b	c	d	e
1.	1.3 +2.6	5 7.6 +3.8	7 4.5 7 +2 1.2	1 8.5 +1 7.3 6	6 7.8 5 7 +2.1 1
2.	4.7 −3.2	6 7.5 −4 3.7	8 7.5 8 −3 4.1	7 5.9 −2 6.6 9	8 7.5 2 −1 2.4 0 3
3.	3.2 4.3 +6.4	5 2.7 −2 6.9	5 3.2 5 1 3.1 +3 1.2 8	6 4.3 −2 5.5 6	1 6.1 0 6 3 4.2 5 +2 1.3
4.	4 7.3 4 −1 3.7 6	1.3 7 2 4.2 3 5 +5.0 5 1	2 5 −1.4 3 5	5.6 0 3 2.7 5 1 +8.8 3 2	3 1.4 2 3 −1 2.8 3
5.	5 7.4 6 3 1.5 9 4 2.3 4 +2.6 7	5.7 0 8 −1.4 3 9	8 3.2 7 5 1 4.2 3 8 8.6 7 5 +3 4.8 7 3	4 8.0 0 3 −1 3.7 4 6	4 7.5 7 8 1 4.4 8 3 7 3.2 4 1 +4 2.9 6 7

Perfect score: 25   My score: _____

9

Problem Solving

Solve each problem.

1. The water level of the lake rose 2.8 feet during March, 4.3 feet during April, and 1.7 feet during May. How much did the water level rise during these three months?

The water level rose _____ feet.

2. In problem 1, how much more did the water level rise during April than during May?

It rose _____ more feet during April.

3. Mr. Tadlock purchased a suit for $97.95 and a topcoat for $87.50. What was the total cost of these articles?

The total cost was $_____.

4. In problem 3, how much more did the suit cost than the topcoat?

The suit cost $_____ more.

5. Last season a certain baseball player had a batting average of .285. This season his batting average is .313. How much has the player's batting average improved?

His average has improved by _____.

6. The thicknesses of three machine parts are .514 centimeter, .317 centimeter, and .178 centimeter. What is the combined thickness of the parts?

The combined thickness is _____ centimeters.

7. Mrs. Dutcher's lot is 60.57 meters long. Mrs. Poole's lot is 54.73 meters long. How much longer is Mrs. Dutcher's lot than Mrs. Poole's lot?

Mrs. Dutcher's lot is _____ meters longer.

8. Mrs. Jolls purchased a dress for $42.95, a pair of shoes for $19, and a purse for $11.49. What was the total amount of these purchases?

The total amount was $_____.

1.	2.
3.	4.
5.	6.
7.	8.

Perfect score: 8   My score: _____

10

Lesson 7   Multiplication

NAME _____

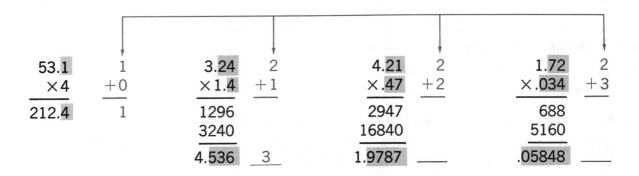

Multiply.

	a	*b*	*c*	*d*	*e*
1.	3 2.5 ×5	4.5 7 ×8	6 7 2 ×.4	1 6 7 8 ×.0 7	8.7 6 5 ×9
2.	3 1.4 ×.0 9	2.2 3 ×.7	4 1 7 ×.0 0 9	.4 1 8 ×.6	1 6 7.8 ×.0 0 8
3.	3.1 4 ×.1 7	6 7.4 ×6.7	5.0 9 ×.0 5 8	.7 2 4 ×.4 6	1 4 8.6 ×2.9
4.	6 5.7 ×.6 4 8	.5 8 4 ×3 5.6	6 9.2 ×4.6 3	7.5 4 ×6 0.7	2.4 0 8 ×5.6 9
5.	3.7 5 ×1.2 4	3.1 4 ×.5 2 6	.9 5 7 ×6.1 8	3.2 7 ×4.3 8	2.1 2 3 ×4.2 5

Perfect score: 25   My score: _____

11

Problem Solving

Solve each problem.

1. Each case of batteries weighs 17.3 kilograms. How much would 6 cases of batteries weigh?

They would weigh _____ kilograms.

2. Six and seventy-five hundredths tons of ore can be processed each hour. At that rate, how many tons of ore can be processed during an 8-hour period?

_____ tons of ore can be processed.

3. An article weighs 6.47 kilograms. What would be the weight of 24 such articles?

The weight would be _____ kilograms.

4. Mr. Swank's car averages 18.7 miles per gallon of gasoline. How many miles would he be able to travel on 12.5 gallons of gasoline?

He would be able to travel _____ miles.

5. An industrial machine uses 4.75 liters of fuel each hour. At that rate how many liters of fuel will be used in 6.5 hours?

_____ liters of fuel will be used.

6. What would be the cost of a 6.2-kilogram roast at $5.40 per kilogram?

The cost would be $_____.

7. Each sheet of paper is .043 centimeter thick. What is the combined thickness of 25 sheets?

It is _____ centimeters.

8. Arlene runs 1.5 kilometers each day. How far will she run in 5 days?

She will run _____ kilometers.

1.	2.
3.	4.
5.	6.
7.	8.

Perfect score: 8   My score: _____

12

Lesson 8   Division

$.73\overline{)21.9}$ ⟶ To get a whole number divisor, multiply both .73 and 21.9 by __100__ .

```
              30                    shorter way
          73)2190                      30
             2190                 .73)21.90
           _____                    21 90
              0                      _____
              0                        0
           _____                      0
              0                      _____
                                       0
```

$.059\overline{).1357}$ ⟶ To get a whole number divisor, multiply both .059 and .1357 by _____ .

```
              2.3                   shorter way
          59)135.7                    2.3
             118 0                .059).135.7
            _____                   118 0
              177                   _____
              177                     177
            _____                    177
              0                     _____
                                      0
```

Divide.

	a	*b*	*c*	*d*
1.	$.61\overline{)3.0\ 5}$	$9.1\overline{)4.5\ 5}$	$.071\overline{).6\ 3\ 9}$	$1.37\overline{).9\ 5\ 9}$
2.	$.37\overline{).9\ 9\ 9}$	$.95\overline{)7\ 6}$	$.026\overline{)1.3\ 7\ 8}$	$16.7\overline{)2.0\ 0\ 4}$
3.	$.03\overline{).7\ 9\ 8}$	$.08\overline{)2\ 0.0\ 8}$	$.47\overline{)9.7\ 2\ 9}$	$25.3\overline{).9\ 2\ 8\ 5\ 1}$

Perfect score: 12   My score: _____

13

Problem Solving

Solve each problem.

1. A rope 40.8 meters long is to be cut into 4 pieces of the same length. How long will each piece be?

Each piece will be _____ meters long.

2. Each can of oil costs $.92. How many cans of oil can be purchased with $23?

_____ cans of oil can be purchased.

3. The cans in a case weigh 9.6 kilograms. Each can weighs .6 kilogram. How many cans are there?

There are _____ cans in the case.

4. Each sheet of paper is .016 centimeter thick. How many sheets will it take to make a stack of paper 18 centimeters high?

It will take _____ sheets.

5. Amy spent $9.60 for meat. A pound of meat sells for $2.40. How many pounds did she buy?

She bought _____ pounds.

6. A machine uses .75 liter of fuel each hour. At that rate, how long will it take to use 22.5 liters of fuel?

It will take _____ hours.

7. Each small transistor weighs .08 gram. How many transistors will it take to weigh 20.4 grams?

It will take _____ transistors.

8. It takes a wheel .6 second to make a revolution. What part of a revolution will the wheel make in .018 second?

It will make _____ of a revolution.

1.	2.
3.	4.
5.	6.
7.	8.

Perfect score: 8   My score: _____

14

Lesson 9   Fractions and Mixed Numerals in Simplest Form

Study how to change a fraction or mixed numeral to simplest form.

Factors of 6
1, 2, 3, 6

Factors of 15
1, 3, 5, 15

$$\frac{6}{15}=\frac{6\div3}{15\div3}$$

Divide 6 and 15 by their greatest common factor.

$$=\frac{2}{5}$$

Factors of 8
1, 2, 4, 8

Factors of 10
1, 2, 5, 10

$$7\frac{8}{10}=7+\frac{8}{10}$$

$$=7+\frac{8\div2}{10\div2}$$

Divide 8 and 10 by their greatest common factor.

$$=7+\frac{4}{5}\text{ or }7\frac{4}{5}$$

$$2\frac{1}{5}$$
$$5\overline{)11}$$
$$\underline{10}$$
$$1$$

$$1\div5=\frac{1}{5}$$

$$5\frac{2}{3}$$
$$3\overline{)17}$$
$$\underline{15}$$
$$2$$

$$2\div3=\frac{2}{3}$$

Write each of the following in simplest form.

	a	*b*	*c*
1.	$\frac{9}{12}$	$2\frac{2}{4}$	$\frac{13}{5}$
2.	$\frac{2}{8}$	$3\frac{6}{10}$	$\frac{19}{3}$
3.	$\frac{10}{16}$	$7\frac{8}{12}$	$\frac{17}{2}$
4.	$\frac{18}{36}$	$5\frac{15}{20}$	$\frac{12}{8}$
5.	$\frac{15}{45}$	$2\frac{12}{28}$	$\frac{16}{10}$

Perfect score: 15   My score: _____

Lesson 10 Renaming Numbers

$\dfrac{3}{4} = \dfrac{\blacksquare}{8}$

| Multiply both the numerator and the denominator by the same number. |

$= \dfrac{3 \times 2}{4 \times 2}$

$= \dfrac{6}{8}$ ⤹ Choose 2 so the new denominator is 8.

$4 = \dfrac{\blacksquare}{8}$

$\dfrac{4}{1} = \dfrac{4 \times 8}{1 \times 8}$

$= \dfrac{32}{8}$ ⤹ Choose 8 so the new denominator is 8.

Name the whole number as a fraction whose denominator is 1.

$9\dfrac{2}{3} = \dfrac{\blacksquare}{3}$

$9\dfrac{2}{3} = \dfrac{(3 \times 9) + 2}{3}$

$= \dfrac{29}{3}$

| Multiply the whole number by the denominator and add the numerator. |

Use the same denominator.

$7\dfrac{1}{5} = \dfrac{\blacksquare}{5}$

$7\dfrac{1}{5} = \dfrac{(5 \times 7) + 1}{5}$

$= \dfrac{36}{5}$

Rename.

	a	*b*	*c*	*d*
1.	$\dfrac{1}{2} = \dfrac{\blacksquare}{6}$	$\dfrac{2}{5} = \dfrac{\blacksquare}{15}$	$\dfrac{3}{8} = \dfrac{\blacksquare}{16}$	$\dfrac{5}{6} = \dfrac{\blacksquare}{12}$
2.	$6 = \dfrac{\blacksquare}{4}$	$3 = \dfrac{\blacksquare}{10}$	$7 = \dfrac{\blacksquare}{3}$	$5 = \dfrac{\blacksquare}{2}$
3.	$4\dfrac{1}{2} = \dfrac{\blacksquare}{2}$	$6\dfrac{3}{4} = \dfrac{\blacksquare}{4}$	$2\dfrac{7}{10} = \dfrac{\blacksquare}{10}$	$3\dfrac{5}{6} = \dfrac{\blacksquare}{6}$
4.	$2 = \dfrac{\blacksquare}{12}$	$\dfrac{9}{10} = \dfrac{\blacksquare}{50}$	$1\dfrac{7}{8} = \dfrac{\blacksquare}{8}$	$\dfrac{4}{5} = \dfrac{\blacksquare}{10}$

Perfect score: 16   My score: _____

Lesson 11 Addition and Subtraction

$$\frac{1}{7}$$
$$\frac{5}{7}$$
$$+\frac{2}{7}$$
$$\frac{8}{7}=1\frac{1}{7}$$

$$1\frac{3}{8} \longrightarrow 1\frac{9}{24}$$
$$2\frac{1}{3} \longrightarrow 2\frac{8}{24}$$
$$+3\frac{1}{2} \longrightarrow +3\frac{12}{24}$$
$$6\frac{29}{24}=\underline{\hspace{2cm}}$$

$$\frac{7}{10}$$
$$-\frac{3}{10}$$
$$\frac{4}{10}=\frac{2}{5}$$

$$3\frac{1}{6} \longrightarrow 3\frac{1}{6} \longrightarrow 2\frac{7}{6}$$
$$-1\frac{1}{2} \longrightarrow -1\frac{3}{6} \longrightarrow -1\frac{3}{6}$$
$$1\frac{4}{6}=\underline{\hspace{2cm}}$$

> To add or subtract fractions having different denominators, rename either or both fractions so they have the same denominator. Then add or subtract.

Write each sum or difference in simplest form.

	a	*b*	*c*	*d*
1.	$\frac{2}{5}$ $+\frac{2}{5}$	$1\frac{4}{7}$ $+\frac{2}{7}$	$\frac{8}{9}$ $-\frac{7}{9}$	$1\frac{7}{10}$ $-\frac{4}{10}$
2.	$\frac{5}{8}$ $+\frac{1}{8}$	$2\frac{5}{6}$ $+3\frac{2}{6}$	$\frac{7}{8}$ $-\frac{3}{8}$	$2\frac{4}{7}$ $-1\frac{4}{7}$
3.	$\frac{2}{3}$ $\frac{1}{4}$ $+\frac{1}{2}$	$\frac{5}{8}$ $\frac{1}{2}$ $+\frac{1}{4}$	$2\frac{1}{3}$ $4\frac{1}{6}$ $+5\frac{1}{4}$	$2\frac{1}{2}$ $4\frac{1}{5}$ $+2\frac{3}{4}$
4.	$\frac{11}{12}$ $-\frac{3}{4}$	$\frac{13}{15}$ $-\frac{2}{3}$	$5\frac{1}{8}$ $-2\frac{2}{3}$	$8\frac{3}{4}$ $-2\frac{11}{12}$

Perfect score: 16   My score: _____

Problem Solving

Solve. Write each answer in simplest form.

1. Kelly purchased $\frac{3}{8}$ pound of cheddar cheese and Jerry purchased $\frac{3}{8}$ pound of brick cheese. How much cheese did they purchase?

They purchased _____ pound of cheese.

2. Doris had $\frac{5}{6}$ yard of material before she used $\frac{1}{6}$ yard. How much material does she have left?

She has _____ yard of material left.

3. One fourth of the house was painted yesterday and $\frac{1}{2}$ was painted today. What fractional part of the house was painted on those two days?

_____ of the house was painted.

4. Owen weighs $114\frac{1}{2}$ pounds and Jesse weighs $107\frac{3}{4}$ pounds. How much more does Owen weigh than Jesse?

Owen weighs _____ pounds more.

5. What is the combined weight of the boys in problem 4?

The combined weight is _____ pounds.

6. Yesterday Gail spent $5\frac{1}{2}$ hours in school, $1\frac{3}{4}$ hours playing, and $1\frac{1}{4}$ hours doing homework. How much time did she spend on these activities?

She spent _____ hours.

7. It is $6\frac{1}{4}$ miles to the beach and $4\frac{1}{2}$ miles to the ball park. How much closer is it to the ball park than to the beach?

It is _____ miles closer to the ball park.

8. In the broad-jump contest, Glenda had a jump of $14\frac{7}{12}$ feet. Horace had a jump of $12\frac{3}{4}$ feet. How much farther did Glenda jump than Horace?

Glenda jumped _____ feet farther.

1.	2.
3.	**4.**
5.	**6.**
7.	**8.**

Perfect score: 8   My score: _____

18

Lesson 12   Multiplication

$$\frac{5}{6} \times \frac{1}{3} = \frac{5 \times 1}{6 \times 3}$$
Multiply numerators.
Multiply denominators.

$$= \frac{5}{18}$$

$$5 \times \frac{3}{4} \times \frac{1}{2} = \frac{5 \times 3 \times 1}{1 \times 4 \times 2}$$

$$= \underline{\qquad}$$

$$4\frac{1}{2} \times 5\frac{2}{3} \times 1\frac{3}{5} = \frac{\overset{3}{\cancel{9}}}{2} \times \frac{17}{\underset{1}{\cancel{3}}} \times \frac{8}{5}$$

$$= \frac{\overset{3}{\cancel{9}}}{\underset{1}{2}} \times \frac{17}{\underset{1}{\cancel{3}}} \times \frac{\overset{4}{\cancel{8}}}{5}$$

Divide a numerator and a denominator by a common factor.

$$= \frac{3 \times 17 \times \underline{\qquad}}{1 \times \underline{\qquad} \times \underline{\qquad}}$$

$$= \underline{\qquad} \text{ or } \underline{\qquad}$$

Write each product in simplest form.

	a	b	c	d
1.	$\frac{3}{5} \times \frac{1}{4}$	$\frac{2}{3} \times \frac{4}{5}$	$\frac{1}{3} \times \frac{1}{5} \times \frac{1}{2}$	$\frac{3}{4} \times \frac{1}{2} \times \frac{3}{5}$
2.	$\frac{3}{10} \times \frac{2}{5}$	$\frac{2}{3} \times \frac{7}{8}$	$\frac{5}{8} \times \frac{3}{10} \times \frac{5}{6}$	$\frac{7}{12} \times \frac{6}{7} \times \frac{2}{3}$
3.	$\frac{3}{4} \times \frac{8}{9}$	$\frac{3}{5} \times \frac{5}{12}$	$\frac{11}{12} \times \frac{3}{4} \times \frac{8}{11}$	$\frac{7}{16} \times \frac{4}{5} \times \frac{5}{8}$
4.	$1\frac{1}{2} \times \frac{5}{7}$	$2\frac{1}{3} \times \frac{5}{12}$	$1\frac{7}{8} \times \frac{2}{3} \times \frac{3}{4}$	$\frac{1}{2} \times 3\frac{1}{3} \times \frac{5}{6}$
5.	$\frac{3}{5} \times 3\frac{1}{3}$	$\frac{5}{6} \times 3\frac{1}{2}$	$1\frac{1}{2} \times 2\frac{1}{3} \times \frac{1}{4}$	$\frac{2}{3} \times 2\frac{1}{2} \times 1\frac{3}{4}$
6.	$2\frac{2}{3} \times 1\frac{3}{4}$	$5\frac{1}{2} \times 3\frac{1}{6}$	$1\frac{2}{3} \times 3\frac{1}{2} \times 2\frac{1}{4}$	$6\frac{1}{2} \times 2\frac{1}{3} \times 1\frac{3}{5}$

Perfect score: 24   My score: _____

Problem Solving

Solve each problem. Write each answer in simplest form.

1. The tank on Mr. Kent's lawn mower will hold $\frac{3}{4}$ gallon of gasoline. Suppose the tank is $\frac{1}{2}$ full. How many gallons of gasoline are in the tank?

_____ gallon of gasoline is in the tank.

2. Band practice lasted $1\frac{1}{4}$ hours. Two thirds of the time was spent marching. How much time was spent marching?

_____ hour was spent marching.

3. An industrial engine uses $2\frac{1}{2}$ gallons of fuel an hour. How many gallons of fuel will be used in operating the engine $1\frac{3}{4}$ hours?

_____ gallons will be used.

4. Stanley has a board which is $3\frac{3}{4}$ feet long. Mike has a board which is $\frac{2}{3}$ as long. How long is Mike's board?

Mike's board is _____ feet long.

5. It is $\frac{1}{4}$ mile around the track. June ran around the track $2\frac{1}{2}$ times. How far did June run?

June ran _____ mile.

6. Steve can broad jump $14\frac{1}{2}$ feet. His younger sister can broad jump $\frac{3}{4}$ as far. How far can his sister broad jump?

His sister can broad jump _____ feet.

7. The football team practices $2\frac{1}{2}$ hours on each of 5 weekday afternoons. How many hours does the team practice each week?

The team practices _____ hours.

8. Mrs. Prell drives $6\frac{1}{2}$ miles each day. How many miles will she drive in 6 days?

She will drive _____ miles.

1.
2.
3.
4.
5.
6.
7.
8.

Perfect score: 8   My score: _____

20

Lesson 13 Division

reciprocals

$$\frac{4}{7} \times \frac{7}{4} = 1$$

reciprocals

$$5 \times \frac{1}{5} = 1$$

reciprocals

$$2\frac{3}{4} \times \frac{4}{11} = 1$$

If two numbers are reciprocals, their product is ____.

Multiply by the reciprocal.

$$\frac{3}{8} \div \frac{4}{5} = \frac{3}{8} \times \frac{5}{4}$$

$$= \frac{15}{32}$$

To divide any number, multiply by its reciprocal.

$$6\frac{1}{2} \div \frac{3}{4} = \frac{13}{2} \times \underline{\quad}$$

$$= \underline{\quad}$$

$$\frac{2}{3} \div 1\frac{1}{2} = \frac{2}{3} \times \underline{\quad}$$

$$= \underline{\quad}$$

Write each quotient in simplest form.

	a	b	c	d
1.	$\frac{1}{2} \div \frac{3}{4}$	$\frac{7}{8} \div \frac{2}{3}$	$\frac{4}{5} \div \frac{4}{7}$	$\frac{5}{8} \div \frac{7}{10}$
2.	$\frac{4}{5} \div 4$	$8 \div \frac{2}{3}$	$\frac{9}{10} \div 3$	$9 \div \frac{3}{5}$
3.	$1\frac{1}{2} \div \frac{2}{3}$	$3\frac{1}{3} \div \frac{5}{6}$	$2\frac{1}{2} \div \frac{7}{10}$	$4\frac{1}{3} \div \frac{7}{8}$
4.	$\frac{7}{8} \div 2\frac{1}{2}$	$\frac{7}{8} \div 1\frac{3}{4}$	$\frac{5}{6} \div 2\frac{2}{3}$	$\frac{3}{4} \div 1\frac{4}{5}$
5.	$2 \div 1\frac{7}{8}$	$4\frac{1}{2} \div 3$	$6 \div 1\frac{1}{8}$	$3\frac{1}{3} \div 5$
6.	$1\frac{1}{2} \div 2\frac{2}{3}$	$3\frac{1}{4} \div 1\frac{7}{8}$	$4\frac{1}{2} \div 1\frac{1}{2}$	$5\frac{1}{4} \div 1\frac{1}{8}$

Perfect score: 24   My score: _____

Problem Solving

Solve. Write each answer in simplest form.

1. Football practice lasted $2\frac{1}{2}$ hours. An equal amount of time was spent on blocking, tackling, passing, and kicking. How much time was spent on each?

_____ hour was spent on each.

2. Three-fourths gallon of gasoline was poured into 4 containers. Each container held the same amount. How much gasoline was poured into each container?

_____ gallon was poured into each container.

3. Suppose a motorboat uses $1\frac{1}{4}$ gallons of fuel each hour. At this rate how long can the boat be operated before 10 gallons of fuel are used?

It can be operated _____ hours.

4. Due to a heavy rain, the water level of the lake was rising $\frac{2}{3}$ foot an hour. At this rate how long will it take the water level to rise $\frac{3}{4}$ foot?

It will take _____ hours.

5. Manny walks $\frac{3}{4}$ mile in going to school and back. How far does Manny live from school?

He lives _____ mile from school.

6. In one hour, $\frac{3}{4}$ pint of a chemical can be filtered. How long will it take to filter $2\frac{1}{2}$ pints of the chemical?

It will take _____ hours.

7. A flight leaves the airport every $1\frac{1}{4}$ minutes. How many flights will leave each hour?

_____ flights will leave each hour.

8. Dolly's mother plans to lose $1\frac{1}{4}$ pounds each week. At this rate how long will it take her to lose 15 pounds?

It will take _____ weeks.

1.
2.
3.
4.
5.
6.
7.
8.

Perfect score: 8   My score: _____

22

CHAPTER 1 TEST

Add or subtract. Write each answer in simplest form.

	a	b	c	d	e

1.
   $\begin{array}{r} 4\ 3\ 4\ 7 \\ 2\ 2\ 1\ 8 \\ +\ 1\ 2\ 5\ 6 \\ \hline \end{array}$
   $\begin{array}{r} 2\ 5.4\ 3 \\ 2.6\ 5 \\ +\ 3\ 4.6\ 7 \\ \hline \end{array}$
   $\begin{array}{r} 5\ 2.1\ 4\ 3 \\ 4.6\ 7\ 5 \\ +\ 1\ 2.5\ 8 \\ \hline \end{array}$
   $\begin{array}{r} \frac{1}{4} \\ \frac{1}{6} \\ +\frac{1}{12} \\ \hline \end{array}$
   $\begin{array}{r} 2\frac{2}{3} \\ 4\frac{3}{4} \\ +5\frac{3}{8} \\ \hline \end{array}$

2.
   $\begin{array}{r} 8\ 3\ 1\ 4 \\ -7\ 8\ 7 \\ \hline \end{array}$
   $\begin{array}{r} 6.7\ 5\ 2 \\ -1.4\ 3\ 8 \\ \hline \end{array}$
   $\begin{array}{r} 1\ 3.0\ 0\ 7 \\ -7.6\ 5\ 9 \\ \hline \end{array}$
   $\begin{array}{r} \frac{3}{4} \\ -\frac{2}{3} \\ \hline \end{array}$
   $\begin{array}{r} 3\frac{1}{3} \\ -1\frac{5}{6} \\ \hline \end{array}$

Multiply or divide. Write each answer in simplest form.

3.
   $\begin{array}{r} 5\ 1\ 3 \\ \times 6\ 7 \\ \hline \end{array}$
   $\begin{array}{r} 1\ 5\ 7\ 2 \\ \times 3\ 4\ 8 \\ \hline \end{array}$
   $\dfrac{2}{3}\times\dfrac{5}{7}$
   $2\times\dfrac{7}{8}$
   $1\dfrac{1}{8}\times\dfrac{1}{3}$

4.
   $\begin{array}{r} 7\ 1.4\ 3 \\ \times .2\ 4 \\ \hline \end{array}$
   $\begin{array}{r} 3\ 6\ 7.1 \\ \times .4\ 1\ 5 \\ \hline \end{array}$
   $3\dfrac{1}{7}\times 2\dfrac{3}{4}$
   $2\div\dfrac{2}{3}$
   $\dfrac{7}{8}\div 2$

5. $28\overline{)5\ 9\ 7}$ $1.25\overline{)4.8\ 7\ 5}$ $1.63\overline{).1\ 9\ 5\ 6}$ $1\dfrac{5}{8}\div\dfrac{3}{4}$ $4\dfrac{1}{2}\div 3\dfrac{1}{3}$

Perfect score: 25   My score: _____

PRE-TEST—Equations

2

Solve each equation.

	a	b	c
1.	$2x = 12$	$5y = 25$	$6z = 96$
2.	$\dfrac{d}{3} = 5$	$\dfrac{e}{6} = 7$	$\dfrac{f}{4} = 13$
3.	$r + 7 = 12$	$s + 3 = 25$	$t + 12 = 20$
4.	$g - 4 = 8$	$h - 5 = 12$	$j - 15 = 15$
5.	$72 = 4m$	$8n = 28 + 28$	$9 = \dfrac{p}{5}$
6.	$18 = a + 6$	$b + 4 = 12 + 3$	$13 = c - 4$
7.	$u - 12 = 23 + 7$	$v + 8 = 8$	$w - 8 = 8$

Perfect score: 21   My score: _____

Lesson 1   Number Phrases

NAME _____

Letters like a, b, n, x, and so on, can be used to stand for numbers.

word phrase	number phrase	
Some number a added to 7	$7+a$	If $a=5$, then $7+a=7+$ _5_ or _12_ .
Some number b decreased by 4	$b-4$	If $b=6$, then $b-4=$ _6_ $-$ 4 or _2_ .
The product of 3 and some number n	$3\times n$ or $3n$	If $n=2$, then $3n=3\times$ _2_ or ___ .
15 divided by some number x	$\frac{15}{x}$ or $15\div x$	If $x=3$, then $15\div x=15\div$ ___ or ___ .

Write a number phrase for each of the following.

              a b

1. Some number c subtracted from 11 _____   Five more than the number b _____

2. A certain number d increased by 12 _____   Some number t divided by 2 _____

3. The product of some number n and 8 _____   Four less than some number x _____

4. Eight divided by some number j _____   The product of $\frac{1}{2}$ and y _____

Complete the following.

5. If $r=3$, then $12-r=$ _____ $-$ _____ or _____ .

6. If $s=9$, then $7+s=$ _____ $+$ _____ or _____ .

7. If $t=3$, then $48\div t=$ _____ $\div$ _____ or _____ .

8. If $u=72$, then $\frac{1}{4}u=$ _____ $\times$ _____ or _____ .

9. If $v=12$, then $4v=$ _____ $\times$ _____ or _____ .

10. If $w=6$, then $w-6=$ _____ $-$ _____ or _____ .

11. If $x=24$, then $\frac{x}{3}=$ _____ $\div$ _____ or _____ .

Perfect score: 29   My score: _____

Lesson 2   Equations

An **equation** like $x+2=9$ states that both $x+2$ and 9 name the same number.

sentence	equation	
The sum of some number and 2 is 9.	$x+2=9$	$x=$ _7_ because _7_ $+2=9$.
Twelve divided by some number is 6.	$12 \div x = 6$ or $\frac{12}{x}=6$	$x=$ _2_ because $12 \div$ ___ $=6$.
Seven decreased by some number is 5.	___ $- x =$ ___	$x=$ ___ because ___ $-$ ___ $=$ ___.

Write an equation for each of the following.

a   b

1. Some number a increased by 6 is 20. _____   A number p decreased by 7 is 15. _____

2. Twenty divided by some number y is 4. _____   One half of a number t is equal to 14. _____

3. The sum of a certain number b and 7 is 14. _____   Twelve more than some number v is 18. _____

4. The product of 2 and some number n is 12. _____   Some number d divided by 3 is equal to 14. _____

Complete the following.

5. $x+8=12$ $x=$ _____ because _____ $+8=12$.

6. $9r=45$ $r=$ _____ because $9 \times$ _____ $=45$.

7. $6=\frac{1}{2}d$ $d=$ _____ because $6=\frac{1}{2} \times$ _____.

8. $b-6=8$ $b=$ _____ because _____ $-6=8$.

9. $w \div 3 = 2$ $w=$ _____ because _____ $\div 3=2$.

10. $e+16=18$ $e=$ _____ because _____ $+16=18$.

11. $35=27+c$ $c=$ _____ because $35=27+$ _____.

Perfect score: 22   My score: _____

Lesson 3   Solving Equations

NAME _____

To solve an equation, you can divide both sides of it by the same non-zero number.

$$4m = 52$$

$$\frac{4m}{4} = \frac{52}{4}$$

$$\frac{\overset{1}{\cancel{4}}m}{\cancel{4}_{1}} = \frac{\overset{13}{\cancel{52}}}{\cancel{4}_{1}}$$

$$m = 13$$

Check

$$4m = 52$$
$$4 \times 13 = 52$$
$$52 = 52$$

To change $4m$ to m,
both sides were divided by _____.

$$13y = 100 - 9$$

$$\frac{13y}{13} = \frac{91}{13}$$

$$\frac{\overset{1}{\cancel{13}}y}{\cancel{13}_{1}} = \frac{\overset{7}{\cancel{91}}}{\cancel{13}_{1}}$$

$$y = \text{\_\_\_\_\_}$$

To change $13y$ to y,
both sides were divided by _____.

Solve each equation.

	a	*b*	*c*
1.	$3w = 12$	$3b = 51$	$8m = 100 - 4$
2.	$72 = 2a$	$54 = 3c$	$96 - 20 = 4r$
3.	$6e = 84$	$25s = 75$	$4d = 75 - 7$
4.	$14x = 42$	$75 = 15m$	$3y = 100 - 28$

Perfect score: 12   My score: _____

27

Problem Solving

Study the first problem. Solve problems **2-5** in a similar way.

1. John bought several model kits for $9 each. He spent $36. How many kits did he buy?

If x stands for the number of kits he bought, then ___9x___ stands for the cost of all the kits.

Equation: ___$9x = 36$___ $x =$ ___4___

John bought ___4___ model kits.

2. A train travels 70 miles per hour. How long does it take for this train to make a 630-mile trip?

If x stands for the number of hours for the trip, then

_____ stands for the total number of miles.

Equation: _____ $x =$ _____

It takes _____ hours to make the trip.

3. Three pounds of apples cost $2.34 (234 cents). How much does 1 pound of apples cost?

If x stands for the cost of 1 pound, then _____ stands for the cost of 3 pounds.

Equation: _____ $x =$ _____

One pound of apples costs _____ cents.

4. Eight loaves of bread cost $7.84 (784 cents). How much does 1 loaf of bread cost?

If x stands for the cost of 1 loaf, then _____ stands for the cost of 8 loaves.

Equation: _____ $x =$ _____

One loaf of bread costs _____ cents.

5. A board is 84 inches long. How many feet long is this board? (There are 12 inches in 1 foot.)

If x stands for the number of feet, then _____ stands for the number of inches.

Equation: _____ $x =$ _____

The board is _____ feet long.

1.

2.

3.

4.

5.

Perfect score: 16   My score: _____

Lesson 4   Solving Equations

To solve an equation, you can multiply both sides of it by the same number.

$\frac{a}{5} = 35$

$5 \times \frac{a}{5} = 5 \times 35$

$\frac{\overset{1}{\cancel{5}} \times a}{\underset{1}{\cancel{5}}} = 175$

$a = 175$

Check

$\frac{a}{5} = 35$

$\frac{175}{5} = 35$

$35 = 35$

To change $\frac{a}{5}$ to a, both
sides were multiplied by _____.

$r \div 3 = 11 + 34$

$(r \div 3) \times 3 = 45 \times _____$

$r = _____$

To change $r \div 3$ to r, both
sides were multiplied by _____.

Solve each equation.

		a	*b*	*c*

1. $\frac{a}{8} = 7$ $\frac{b}{13} = 9$ $\frac{c}{4} = 6 + 12$

2. $16 = \frac{r}{8}$ $8 = s \div 7$ $2 \times 9 = \frac{t}{5}$

3. $g \div 17 = 5$ $15 = \frac{h}{5}$ $7 \times 6 = \frac{j}{3}$

4. $\frac{m}{15} = 17$ $23 = \frac{n}{28}$ $p \div 19 = 3 \times 9$

Perfect score: 12   My score: _____

Problem Solving

Study the first problem. Solve problems **2-5** in a similar way.

1. Joe has $\frac{1}{4}$ the number of points he needs to win. He has 36 points. How many points does he need to win?

If x stands for the number of points needed to win, then $\frac{1}{4}x$ or $\frac{x}{4}$ stands for the points he has now.

Equation: _____ $\frac{1}{4}x=36$ or $\frac{x}{4}=36$ _____ $x=$ _____ 144 _____

Joe needs _____ 144 _____ points to win.

2. Mia has $\frac{1}{3}$ the number of points she needs to win. She has 48 points. How many points does she need to win?

If x stands for the total number of points needed to

win, then _____ stands for the points she has now.

Equation: _____ $x=$ _____

Mia needs _____ points to win.

3. Three students are absent. This is $\frac{1}{6}$ of the entire class. How many students are in the class?

If x stands for the total number of students, then

_____ stands for the number of students absent.

Equation: _____ $x=$ _____

There are _____ students in the class.

4. Jim drove 120 miles and stopped for lunch. He had then traveled $\frac{1}{3}$ the total distance of his trip. What is the total distance of his trip?

If x stands for the total distance, then _____ stands for the distance he has already traveled.

Equation: _____ $x=$ _____

The total distance of the trip is _____ miles.

5. Mary worked 12 problems. This was $\frac{1}{5}$ of all she has to work. How many problems does she have to work?

Equation: _____ $x=$ _____

She has _____ problems to work in all.

1.

2.

3.

4.

5.

Perfect score: 15   My score: _____

30

Lesson 5   Solving Equations

To solve an equation, you can subtract the same number from both sides of it.

$v+18=47$

$v+18-18=47-18$

$v+0=29$

$v=29$

Check

$v+18=47$

$29+18=47$

$47=47$

$c+6=43+8$

$c+6-$ _____ $=51-$ _____

$c+$ _____ $=$ _____

$c=$ _____

To change $v+18$ to v, _____ was subtracted from both sides.

To change $c+6$ to c, _____ was subtracted from both sides.

Solve each equation.

	a	*b*	*c*
1.	$d+12=48$	$36+e=84$	$f+14=18+18$
2.	$38=j+13$	$27=9+h$	$20+34=27+l$
3.	$12+w=76$	$114=x+38$	$300-30=y+50$
4.	$200+50=a+212$	$27+b=170+3$	$100-2=c+43$

Perfect score: 12   My score: _____

Problem Solving

Study the first problem. Solve problems **2-5** in a similar way.

1. A rectangle is 8 feet longer than it is wide. If its length is 17 feet, what is its width?

If x stands for the number of feet wide, then $\underline{\quad x+8 \quad}$ stands for the number of feet long.

Equation: $\underline{\quad x+8=17 \quad}$ $\qquad$ $x = \underline{\quad 9 \quad}$

The width of the rectangle is $\underline{\quad 9 \quad}$ feet.

1.

2. A rectangle is 27 inches longer than it is wide. If its length is 45 inches, what is its width?

If x stands for the number of inches wide, then

$\underline{\hspace{3cm}}$ stands for the number of inches long.

Equation: $\underline{\hspace{3cm}}$ $\qquad$ $x = \underline{\hspace{2cm}}$

The width of the rectangle is $\underline{\hspace{2cm}}$ inches.

2.

3. Helen's score of 94 is 8 points higher than Jane's score. What is Jane's score?

If x stands for Jane's score, then $\underline{\hspace{2cm}}$ stands for Helen's score.

Equation: $\underline{\hspace{3cm}}$ $\qquad$ $x = \underline{\hspace{2cm}}$

Jane's score is $\underline{\hspace{2cm}}$.

3.

4. The 17 men at work outnumber the women by 5. How many women are at work?

If x stands for the number of women at work, then

$\underline{\hspace{3cm}}$ stands for the number of men at work.

Equation: $\underline{\hspace{3cm}}$ $\qquad$ $x = \underline{\hspace{2cm}}$

There are $\underline{\hspace{2cm}}$ women at work.

4.

5. The 48-minute trip to work was 19 minutes longer than the trip home from work. How long did it take for the trip home?

If x stands for the number of minutes for the trip home, then $\underline{\hspace{2cm}}$ stands for the trip to work.

Equation: $\underline{\hspace{3cm}}$ $\qquad$ $x = \underline{\hspace{2cm}}$

The trip home took $\underline{\hspace{2cm}}$ minutes.

5.

Perfect score: 16   My score: $\underline{\hspace{2cm}}$

Lesson 6   Solving Equations

To solve an equation, you can add the same number to both sides of it.

$t-3=15$

$t-3+3=15+3$

$t+0=18$

$t=18$

Check

$t-3=15$

$18-3=15$

$15=15$

$b-12=14+3$

$b-12+$ _____ $=17+$ _____

$b+$ _____ $=$ _____

$b=$ _____

To change $t-3$ to t, _____ was added to both sides.

To change $b-12$ to b, _____ was added to both sides.

Solve each equation.

	a	b	c
1.	$b-8=15$	$x-14=36$	$c-3=28+4$
2.	$42=r-12$	$80=e-26$	$20+11=f-14$
3.	$163=a-27$	$9\times9=m-38$	$t-28=102$
4.	$117=w-83$	$200-25=g-83$	$h-75=100+56$

Perfect score: 12   My score: _____

Problem Solving

Study the first problem. Solve problems **2-5** in a similar way.

1. The temperature has fallen 12 degrees since noon. The present temperature is 57 degrees. What was the noon temperature?

If x stands for the noon temperature, then ___$x-12$___ stands for the present temperature.

Equation: ___$x-12=57$___ $x=$ ___69___

The noon temperature was ___69___ degrees.

2. The temperature has fallen 7 degrees since noon. The present temperature is 78 degrees. What was the noon temperature?

If x stands for the noon temperature, then _____ stands for the present temperature.

Equation: _____ $x=$ _____

The noon temperature was _____ degrees.

3. After selling 324 papers, Mr. Merk had 126 papers left. How many papers did he start with?

If x stands for the number of papers he started with,

then _____ stands for the number left.

Equation: _____ $x=$ _____

He had _____ papers to start with.

4. Gene sold his football for \$15.50 (1550 cents). This was 95 cents less than the original cost. What was the original cost?

If x stands for the original cost, then _____ stands for the amount he sold the football for.

Equation: _____ $x=$ _____

The original cost of the football was \$_____.

5. The width of a rectangle is 37 inches shorter than its length. The width is 75 inches. How long is the rectangle?

If x stands for the measure of the length, then

_____ stands for the measure of the width.

Equation: _____ $x=$ _____

The rectangle is _____ inches long.

1.

2.

3.

4.

5.

Perfect score: 16   My score: _____

34

Lesson 7   Solving Equations

Solve each equation.

	a	*b*	*c*

1. $4b = 30 + 30$ $13 + 26 = 3u$ $7v = 42 + 42$

2. $\dfrac{d}{5} = 100 - 40$ $10 - 3 = \dfrac{y}{32}$ $\dfrac{k}{37} = 10 - 8$

3. $g + 27 = 49 - 4$ $100 - 7 = 39 + x$ $43 = n + 12$

4. $p - 6 = 3 + 10$ $56 - 3 = k - 42$ $w - 39 = 90 + 3$

5. $x + 16 = 33 + 12$ $\dfrac{m}{14} = 14 - 4$ $12 + n = 56 + 8$

6. $7 \times 6 = 3t$ $d + 291 = 400 + 26$ $73 - 8 = n + 5$

7. $4 + 8 = \dfrac{q}{12}$ $g - 27 = 2 \times 45$ $\dfrac{x}{15} = 20 - 5$

Perfect score: 21   My score: _____

Problem Solving

Safety Program	
Name	*Points Earned*
Tom	48
Susan	
Bob	
Marilyn	
Al	

The bulletin-board chart was torn and some information is missing. Help complete the chart by using the information in the following problems. Write an equation for each problem. Solve the equation. Answer the problem.

1. Tom has earned 3 times as many points as Susan. How many points has Susan earned?

Equation: _____ $x =$ _____

Susan has earned _____ points.

2. Tom has $\frac{1}{3}$ the number of points that Bob has. How many points does Bob have?

Equation: _____ $x =$ _____

Bob has _____ points.

3. The number of points that Tom has is 27 less than the number of points that Marilyn has. How many points does Marilyn have?

Equation: _____ $x =$ _____

Marilyn has _____ points.

4. The number of points that Tom has earned is 27 more than the number of points that Al has earned. How many points has Al earned?

Equation: _____ $x =$ _____

Al has earned _____ points.

1.

2.

3.

4.

Perfect score: 12   My score: _____

36

CHAPTER 2 TEST

Solve each equation.

	a	b	c
1.	$4m = 40$	$90 = 6n$	$42 - 20 = 2p$
2.	$\dfrac{r}{6} = 7$	$15 = \dfrac{s}{13}$	$\dfrac{t}{2} = 40 + 3$
3.	$a + 9 = 36$	$27 = 6 + b$	$14 + 20 = c + 4$
4.	$x - 9 = 27$	$36 = y - 14$	$z - 6 = 30 + 12$
5.	$42 + 18 = w + 20$	$72 + 18 = \dfrac{x}{6}$	$37 + 12 = y - 18$
6.	$12d = 144$	$17 = \dfrac{e}{3}$	$30m = 3 \times 60$

Write an equation for the problem. Solve.

7. Five workers were absent today. This is $\frac{1}{4}$ of all workers. How many workers are there?

Equation: _____

There are _____ workers.

Perfect score: 20   My score: _____

37

PRE-TEST—Using Equations to Solve Problems

Complete the following.

a	b	c

1. $7x + 2x = $ _____ $9y + y = $ _____ $z + 2z = $ _____

2. $6a - 2a = $ _____ $5b - b = $ _____ $c + 2c = $ _____

Solve each equation.

3. $3r + r = 36$ $5s + s = 42$ $t + 3t = 52$

4. $d + d + 8 = 48$ $e + e + 6 = 74$ $f + f - 5 = 95$

5. $u + 2u + 1 = 10$ $v + 3v + 4 = 24$ $w + 5w + 2 = 50$

Solve each problem.

6. Julie made four times as many widgets as Carmen. They made a total of 60 widgets. How many widgets did Carmen make?

6.

Carmen made _____ widgets.

7. A car averages 72 kilometers per hour. At that rate, how far can the car travel in 3 hours?

7.

The car will travel _____ kilometers.

Perfect score: 17   My score: _____

Lesson 1   Combining Terms

$$3a + 2a = a + a + a + a + a$$
$$= 5a$$

$$3b - 2b = b + b + b - b - b$$
$$= \underline{\ 1b\ } \text{ or } \underline{\ b\ }$$

$$3a + 2a = (3 + 2)a$$
$$= 5a$$

$$3b - 2b = (3 - 2)b$$
$$= \underline{\hspace{1.5cm}} \text{ or } \underline{\hspace{1.5cm}}$$

Complete the following.

	a		b		c
1.	$d + 3d = \underline{\ 4d\ }$		$5e + 2e = \underline{\hspace{1cm}}$		$7f + 2f = \underline{\hspace{1cm}}$
2.	$4g - 3g = \underline{\hspace{1cm}}$		$8h - 4h = \underline{\hspace{1cm}}$		$5j - j = \underline{\hspace{1cm}}$
3.	$2k + k = \underline{\hspace{1cm}}$		$5l - 3l = \underline{\hspace{1cm}}$		$3m + 2m = \underline{\hspace{1cm}}$
4.	$5n + 3n = \underline{\hspace{1cm}}$		$2p - p = \underline{\hspace{1cm}}$		$4q - q = \underline{\hspace{1cm}}$
5.	$8r - 2r = \underline{\hspace{1cm}}$		$5s + 4s = \underline{\hspace{1cm}}$		$5t + t = \underline{\hspace{1cm}}$
6.	$4u + 3u = \underline{\hspace{1cm}}$		$9v - v = \underline{\hspace{1cm}}$		$3w + w = \underline{\hspace{1cm}}$

Complete the following.

	a		b
7.	If $a = 5$, then $3a + 2a = \underline{\ 25\ }$.		If $b = 3$, then $5b - 2b = \underline{\hspace{1cm}}$.
8.	If $c = 2$, then $3c + c = \underline{\hspace{1cm}}$.		If $d = 1$, then $3d - d = \underline{\hspace{1cm}}$.
9.	If $e = 5$, then $2e + 2e = \underline{\hspace{1cm}}$.		If $f = 4$, then $5f - 4f = \underline{\hspace{1cm}}$.
10.	If $g = 2$, then $g + 3g = \underline{\hspace{1cm}}$.		If $h = 5$, then $2h - h = \underline{\hspace{1cm}}$.
11.	If $j = 3$, then $2j + 4j = \underline{\hspace{1cm}}$.		If $k = 3$, then $4k - 3k = \underline{\hspace{1cm}}$.
12.	If $l = 5$, then $3l + 3l = \underline{\hspace{1cm}}$.		If $m = 1$, then $6m - m = \underline{\hspace{1cm}}$.

Perfect score: 28   My score: _____

Lesson 2   Solving Equations

$$x + 5x = 18$$
$$6x = 18$$
$$x = \frac{18}{6}$$
$$x = 3$$

Check

$$x + 5x = 18$$
$$3 + (5 \times 3) = 18$$
$$3 + 15 = 18$$
$$18 = 18$$

$$y + y + 3 = 27$$
$$2y + 3 = 27$$
$$2y = 27 - 3$$
$$2y = 24$$
$$y = 24 \div 2$$
$$y = 12$$

Check

$$y + y + 3 = 27$$
$$12 + 12 + 3 = 27$$
$$27 = 27$$

If $x + 5x = 18$, then $x =$ _____ and $5x =$ _____.

If $y + y + 3 = 27$, then $y =$ _____.

Solve each equation.

	a	*b*	*c*
1.	$4a + a = 25$	$7b + b = 72$	$c + 6c = 49$
2.	$d + d + 2 = 22$	$e + e + 8 = 28$	$f + f - 6 = 30$
3.	$3g + g = 48$	$h + h - 5 = 25$	$5j + j = 54$
4.	$k + k + 4 = 44$	$3l + l = 72$	$m + m - 7 = 19$
5.	$n + 8n = 108$	$p + p + 12 = 60$	$2q + q = 72$

Perfect score: 15   My score: _____

Lesson 3   Problem Solving

Larry is twice as old as Marvin. Their combined age is 24 years. How old is each boy?

If x stands for Marvin's age, then
__2x__ stands for Larry's age.

Equation: _____ $x + 2x = 24$ _____

Marvin is ____8____ years old.

Larry is ____16____ years old.

$$x + 2x = 24$$
$$3x = 24$$
$$x = 24 \div 3$$
$$x = 8$$

Since $x = 8$,
$2x = 2 \times 8$ or 16.

Check
$$x + 2x = 24$$
$$8 + (2 \times 8) = 24$$
$$8 + 16 = 24$$
$$24 = 24$$

Write an equation for each problem. Solve each problem.

1. An office has 28 workers. There are three times as many men as women. How many women are there? How many men are there?

Equation: _____

There are _____ women and _____ men.

1.

2. During the summer Irwin worked 4 times as many days as Mary. They worked a total of 75 days. How many days did each work?

Equation: _____

Mary worked _____ days. Irwin worked _____ days.

2.

3. A truck weighs 4,200 pounds. The weight of the truck body is 6 times that of the engine. How much does the engine weigh? How much does the truck body weigh?

Equation: _____

The engine weighs _____ pounds and the truck

body weighs _____ pounds.

3.

4. Jack is 3 times as old as Lil. Their combined age is 52. How old is each person?

Equation: _____

Lil is _____ years old. Jack is _____ years old.

4.

Perfect score: 12   My score: _____

Lesson 4   Problem Solving

In an election between two girls, 75 votes were cast. Mary received 5 more votes than Jane. How many votes did each girl receive?

If x stands for the number of votes for Jane, then __$x+5$__ stands for the number of votes for Mary.

Equation: __$x+(x+5)=75$__

$$x+(x+5)=75$$
$$2x+5=75$$
$$2x=75-5$$
$$2x=70$$
$$x=35$$

Check
$$x+(x+5)=75$$
$$35+35+5=75$$
$$75=75$$

Jane received __35__ votes.

Since $x=35$,
$x+5=35+5$ or 40.

Mary received __40__ votes.

Write an equation for each problem. Solve each problem.

1. Paul made 7 more gadgets than Gene. Together they made 55 gadgets. How many did each man make?

Equation: _____

Paul made _____ gadgets and Gene made _____.

1.

2. Two pairs of shoes cost $58. One pair costs $6 more than the other. How much did each pair cost?

Equation: _____

One pair cost $ _____ and the other cost $ _____.

2.

3. Sally weighs 16 pounds more than Sue. Their combined weight is 184 pounds. How much does each girl weigh?

Equation: _____

Sue weighs _____ pounds and Sally _____ pounds.

3.

4. Peg has 12 more cases to unload than Mike. They have a total of 150 cases to unload. How many cases does each have to unload?

Equation: _____

Mike has _____ cases and Peg has _____ cases.

4.

Perfect score: 12   My score: _____

Lesson 5   Problem Solving

Max has two boards that have a combined length of 16 feet. One board is 1 foot longer than twice the length of the other. What is the length of each board?

If x stands for the length of the shorter board, then $\underline{2x+1}$ stands for the length of the longer board.

Equation: _____ $x+(2x+1)=16$ _____

The shorter board is __5__ feet long.

The longer board is __11__ feet long.

$x+(2x+1)=16$
$3x+1=16$
$3x=16-1$
$3x=15$
$x=5$

Since $x=5$,
$2x+1=(2\times5)+1$ or 11.

Check
$x+(2x+1)=16$
$5+(2\times5)+1=16$
$5+10+1=16$
$16=16$

Write an equation for each problem. Solve each problem.

1. Mark and Bill have a combined weight of 170 pounds. Mark weighs 40 pounds less than twice Bill's weight. How much does each boy weigh?

Equation: _____

Bill weighs _____ pounds.

Mark weighs _____ pounds.

1.

2. Mary and Betty have saved $43. Betty has saved $3 more than 3 times the amount Mary has saved. How much money has each girl saved?

Equation: _____

Mary has saved $_____.

Betty has saved $_____.

2.

3. A carpenter cut a board that was 10 feet long into two pieces. The longer piece is 2 feet longer than three times the length of the shorter piece. What is the length of each piece?

Equation: _____

The shorter piece is _____ feet long.

The longer piece is _____ feet long.

3.

Perfect score: 9   My score: _____

Lesson 5   Problem Solving

Write an equation for each problem. Solve each problem.

1. Mary said that box A is 2 kilograms heavier than box D. She also said that together these boxes weigh 16 kilograms. How much does each box weigh?

Equation: _____

Box A weighs _____ kilograms.

Box D weighs _____ kilograms.

2. Box C is twice as heavy as box A. Together they weigh 27 kilograms. How much does each box weigh?

Equation: _____

Box A weighs _____ kilograms.

Box C weighs _____ kilograms.

3. Box B weighs 1 kilogram more than twice the weight of box D. They have a combined weight of 22 kilograms. How much does each box weigh?

Equation: _____

Box B weighs _____ kilograms.

Box D weighs _____ kilograms.

4. Mary weighs 1 kilogram more than Mark. Their total weight is 97 kilograms. How much does each person weigh?

Equation: _____

Mark weighs _____ kilograms.

Mary weighs _____ kilograms.

1.

2.

3.

4.

Perfect score: 12   My score: _____

44

Lesson 6   Problem Solving

$$\text{distance} = \text{rate} \times \text{time}$$
$$d = r \times t$$

A robin flew 171 kilometers in 3 hours. At what speed did the robin fly?

$d = r \times t$
$171 = r \times 3$
$\frac{171}{3} = r$
$57 = r$

Equation: _____ $171 = r \times 3$ _____

The robin flew ___57___ kilometers per hour.

Write an equation for each problem. Solve each problem.

1. At 450 kilometers per hour, how far can a plane fly in 5 hours?

Equation: _____

The plane can fly _____ kilometers in 5 hours.

1.

2. The Willards want to travel 744 kilometers in 12 hours. They plan to travel the same distance each hour. At what speed would they travel?

Equation: _____

They would travel _____ kilometers per hour.

2.

3. A ship averages 25 miles per hour. How far can the ship travel in 2 days?

Equation: _____

The ship can travel _____ miles in 2 days.

3.

4. At what speed would a plane have to fly in order to travel 780 kilometers in 2 hours?

Equation: _____

It would fly at _____ kilometers per hour.

4.

5. At 204.8 kilometers per hour, how far can a race car travel in 4 hours?

Equation: _____

It can travel _____ kilometers in 4 hours.

5.

Perfect score: 10   My score: _____

Problem Solving

For all levers, $w \times d = W \times D$.

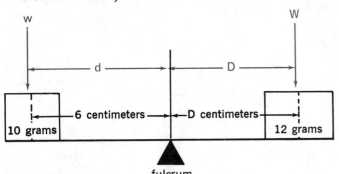

6 centimeters — D centimeters

10 grams   12 grams

fulcrum

To balance the lever (or scale), how far from the fulcrum must the 12-gram weight be placed?

$$w \times d = W \times D$$
$$10 \times 6 = 12 \times D$$
$$\frac{60}{12} = D$$
$$5 = D$$

Check

$$w \times d = W \times D$$
$$10 \times 6 = 12 \times 5$$
$$60 = 60$$

The 12-gram weight must be placed _____ centimeters from the fulcrum.

Write an equation for each problem. Solve each problem.

1. A 60-kilogram boy sits 2 meters from the fulcrum of a seesaw. How far from the fulcrum should a 40-kilogram girl sit so the seesaw is balanced?

Equation: _____

She should sit _____ meters from the fulcrum.

1.

2. How much weight would have to be applied at point A so that the lever is balanced?

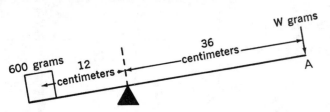

600 grams   12 centimeters   36 centimeters   W grams   A

Equation: _____

_____ grams would have to be applied at point A.

2.

3. What weight is needed at point S on the scale so that the scale is balanced?

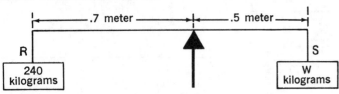

.7 meter   .5 meter

R   S

240 kilograms   W kilograms

Equation: _____

_____ kilograms are needed at point S.

3.

Perfect score: 6   My score: _____

46

CHAPTER 3 TEST

Solve each problem.

1. Marge worked 3 times as many hours as Sue. They worked a total of 32 hours. How many hours did Sue work?

Equation: _____

Sue worked _____ hours.

1.

2. An eraser and a pencil cost 87 cents. The pencil cost 9 cents more than the eraser. How much did the pencil cost?

Equation: _____

The pencil cost _____ cents.

2.

3. Joe and Jim earned 215 points in a contest. Joe earned 5 more than twice as many points as Jim. How many points did each boy earn in the contest?

Equation: _____

Jim earned _____ points.

Joe earned _____ points.

3.

4. Darla scored twice as many points as the combined scores of Mary and Ken. Darla scored 88 points. Mary scored 20 points. How many points did Ken score?

Ken scored _____ points.

4.

5. At 51 miles per hour, how far can a car travel in 3 hours?

It will travel _____ miles in 3 hours.

5.

6. To balance the lever, how far from the fulcrum must the 40-kilogram weight be placed?

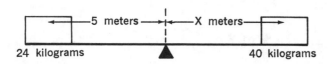

It must be _____ meters from the fulcrum.

6.

Perfect score: 10   My score: _____

47

PRE-TEST—Ratio, Proportion, and Percent

Draw a ring around each proportion below.

$$a \qquad\qquad b$$

1. $\dfrac{3}{16} = \dfrac{6}{24}$ $\dfrac{7}{8} = \dfrac{28}{32}$

2. $\dfrac{8}{20} = \dfrac{4}{5}$ $\dfrac{2}{3} = \dfrac{10}{15}$

3. $\dfrac{7}{9} = \dfrac{21}{27}$ $\dfrac{24}{15} = \dfrac{8}{5}$

Solve each of the following.

4. $\dfrac{n}{3} = \dfrac{9}{27}$ $\dfrac{3}{5} = \dfrac{15}{n}$

5. $\dfrac{5}{6} = \dfrac{n}{36}$ $\dfrac{n}{8} = \dfrac{3}{6}$

6. $\dfrac{8}{24} = \dfrac{n}{15}$ $\dfrac{n}{10} = \dfrac{9}{15}$

7. $\dfrac{10}{25} = \dfrac{8}{n}$ $\dfrac{42}{n} = \dfrac{3}{4}$

Complete the following.

$$a \qquad\qquad\qquad b$$

8. _____ is 12% of 36.   7 is _____% of 16.

9. $\dfrac{1}{2}$ is 50% of _____.   45 is 75% of _____.

10. $\dfrac{2}{5}$ is _____% of $\dfrac{1}{2}$.   _____ is 30% of 200.

11. 3.6 is 80% of _____.   1.8 is _____% of 2.4.

12. _____ is 6.7% of 83.   135 is _____% of 90.

Perfect score: 24   My score: _____

Lesson 1   Ratio

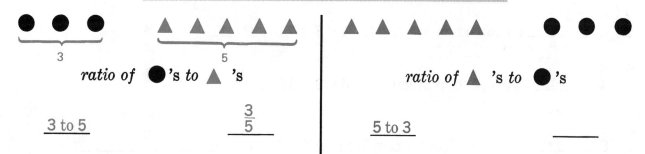

A **ratio** is a comparison of the numbers of two sets.

ratio of ●*'s to* ▲*'s*   *ratio of* ▲*'s to* ●*'s*

3 to 5 $\frac{3}{5}$ 5 to 3   _____

Express the ratio of the number of the first set to the number of the second set in two ways as shown.

			a	*b*
1.	{○,□}	{*,△,□}	2 to 3	$\frac{2}{3}$
2.	{Jim,John}	{Jo,Sue,Ann,Kay}	_____	_____
3.	{1,2,3,4}	{a,b,c}	_____	_____
4.	{Bob,Dick,Al}	{1st,2nd,3rd}	_____	_____
5.	{m,n,o,p,q}	{w,x,y,z}	_____	_____

Express each of the following as a ratio in two ways as shown.

		a	*b*
6.	7 runs in 9 innings	7 to 9	$\frac{7}{9}$
7.	5 boys to 6 girls	_____	_____
8.	3 teachers for 72 students	_____	_____
9.	5 pages in 20 minutes	_____	_____
10.	5 touchdowns in 4 games	_____	_____
11.	11 chairs to 2 tables	_____	_____
12.	6 goals for 9 shots	_____	_____

Perfect score: 20   My score: _____

Lesson 2   Proportions

A **proportion** expresses the equality of two ratios.

$\frac{2}{3} = \frac{4}{6}$ _____is_____ a proportion because $2 \times 6 = 3 \times 4$ is _____true_____ .

$\frac{5}{8} = \frac{3}{4}$ ____is not____ a proportion because $5 \times 4 = 8 \times 3$ is _____false_____ .

$\frac{4}{5} = \frac{7}{8}$ _____ a proportion because $4 \times 8 = 5 \times 7$ is _____ .

$\frac{3}{4} = \frac{9}{12}$ _____ a proportion because $3 \times 12 = 4 \times 9$ is _____ .

Draw a ring around each proportion below.

	a	b
1.	$\frac{2}{3} = \frac{8}{12}$	$\frac{1}{4} = \frac{2}{9}$
2.	$\frac{5}{8} = \frac{11}{16}$	$\frac{5}{6} = \frac{20}{24}$
3.	$\frac{1}{6} = \frac{2}{12}$	$\frac{14}{16} = \frac{7}{8}$
4.	$\frac{5}{12} = \frac{15}{36}$	$\frac{8}{3} = \frac{15}{6}$
5.	$\frac{7}{20} = \frac{15}{40}$	$\frac{9}{24} = \frac{1}{3}$
6.	$\frac{1}{3} = \frac{6}{18}$	$\frac{15}{24} = \frac{5}{8}$
7.	$\frac{7}{12} = \frac{9}{16}$	$\frac{4}{5} = \frac{10}{12}$
8.	$\frac{9}{10} = \frac{90}{100}$	$\frac{8}{10} = \frac{4}{5}$
9.	$\frac{4}{12} = \frac{5}{16}$	$\frac{4}{3} = \frac{5}{4}$
10.	$\frac{12}{25} = \frac{48}{100}$	$\frac{125}{1000} = \frac{1}{8}$

Perfect score: 20   My score: _____

Lesson 3   Proportions

Study how the proportions below are solved.

$$\frac{5}{8}=\frac{15}{n}$$

$5 \times n = 8 \times 15$

$5n = 120$

$n = \underline{\quad 24 \quad}$

$$\frac{2}{3}=\frac{n}{24}$$

$2 \times 24 = 3 \times n$

$48 = 3n$

$\underline{\quad\quad} = n$

$$\frac{5}{n}=\frac{6}{24}$$

$5 \times 24 = n \times 6$

$120 = 6n$

$\underline{\quad\quad} = n$

$$\frac{n}{6}=\frac{20}{24}$$

$n \times 24 = 6 \times 20$

$24n = 120$

$n = \underline{\quad\quad}$

Solve each of the following.

	a	b	c
1.	$\frac{2}{3}=\frac{n}{18}$	$\frac{3}{5}=\frac{n}{25}$	$\frac{3}{4}=\frac{n}{100}$
2.	$\frac{1}{4}=\frac{2}{n}$	$\frac{5}{6}=\frac{10}{n}$	$\frac{7}{8}=\frac{42}{n}$
3.	$\frac{n}{6}=\frac{2}{3}$	$\frac{n}{8}=\frac{21}{24}$	$\frac{n}{3}=\frac{24}{36}$
4.	$\frac{8}{n}=\frac{1}{2}$	$\frac{5}{n}=\frac{20}{28}$	$\frac{4}{n}=\frac{80}{100}$
5.	$\frac{n}{2}=\frac{12}{8}$	$\frac{5}{8}=\frac{n}{1000}$	$\frac{3}{4}=\frac{36}{n}$

Perfect score: 15   My score: _____

Lesson 4 Proportions

A train can travel 120 miles in 2 hours. At that rate, how far can the train travel in 3 hours?

Let n represent the number of miles traveled in 3 hours. Then the following proportions can be obtained by thinking as follows.

Compare the number of hours to the number of miles traveled.	Compare the number of miles traveled to the number of hours.	Compare the first number of hours to the second and the first number of miles to the second.	Compare the second number of hours to the first and the second number of miles to the first.
$\dfrac{2}{120} = \dfrac{3}{n}$	$\dfrac{120}{2} = \dfrac{n}{3}$	$\dfrac{2}{3} = \dfrac{120}{n}$	$\dfrac{3}{2} = \dfrac{n}{120}$
$2n = 360$	$360 = 2n$	$2n = 360$	$360 = 2n$
$n = \underline{\hspace{1cm}}$	$\underline{\hspace{1cm}} = n$	$n = \underline{\hspace{1cm}}$	$\underline{\hspace{1cm}} = n$

Use a proportion to solve each problem.

1. If 8 cases of merchandise cost $60, what would a dozen cases cost?

A dozen cases would cost $_____.

2. Two pounds of apples can be purchased for 98¢. At this rate, what would 1 pound of apples cost?

1 pound of apples would cost _____ ¢.

3. James delivered 450 circulars in 3 hours. At this rate, how many circulars can he deliver in 4 hours?

He can deliver _____ circulars in 4 hours.

4. In his last game the Rams' quarterback threw 18 passes and completed 10. At this rate, how many passes will he complete if he throws 27 passes in a game?

He will complete _____ passes.

5. Mrs. Svage used 3 gallons of paint to cover 1,350 square feet. At this rate, how much paint will be needed to cover 1,800 square feet?

_____ gallons will be needed.

1.

2.

3.

4.

5.

Perfect score: 5   My score: _____

Lesson 5   Problem Solving

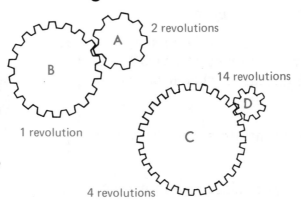

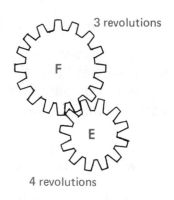

As **A** revolves twice, **B** revolves once.

As **C** revolves 4 times, **D** revolves 14 times.

As **E** revolves 4 times, **F** revolves 3 times.

Use a proportion to solve each problem.

1. When gear **A** has completed 6 revolutions, how many revolutions will gear **B** have made?

Gear **B** will have made _____ revolutions.

2. While gear **B** is making 30 revolutions, how many revolutions will gear **A** make?

Gear **A** will make _____ revolutions.

3. When gear **C** has completed 12 revolutions, how many revolutions will gear **D** have made?

Gear **D** will have made _____ revolutions.

4. While gear **D** is making 84 revolutions, how many revolutions will gear **C** make?

Gear **C** will make _____ revolutions.

5. When gear **E** has completed 56 revolutions, how many revolutions will gear **F** have made?

Gear **F** will have made _____ revolutions.

6. While gear **F** is making 90 revolutions, how many revolutions will gear **E** make?

Gear **E** will make _____ revolutions.

7. When gear **E** has completed 76 revolutions, how many revolutions will gear **F** have made?

Gear **F** will have made _____ revolutions.

| 1. |
| 2. |
| 3. |
| 4. |
| 5. |
| 6. |
| 7. |

Perfect score: 7   My score: _____

Problem Solving

Use a proportion to solve each problem.

1. An orange-juice concentrate is to be mixed with water so that the ratio of water to concentrate is 3 to 1. At this rate, how much concentrate should be mixed with 6 pints of water?

_____ pints of concentrate should be mixed with 6 pints of water.

2. A shoe store sells 4 pairs of black shoes for every 7 pairs of brown shoes. There were 4,900 pairs of brown shoes sold last year. How many pairs of black shoes were sold?

_____ pairs of black shoes were sold.

3. At the Kolbus Building, 3 out of every 7 employees use public transportation. There are 9,800 employees at the building. How many use public transportation?

_____ use public transportation in getting to work.

4. The ratio of box seats at the ball park to general-admission seats is 2 to 7. There are 2,500 box seats. How many general-admission seats are there?

There are _____ general-admission seats.

5. At the snack bar, 7 hot dogs are sold for every 10 hamburgers sold. At this rate, how many hot dogs will be sold if 90 hamburgers are sold?

_____ hot dogs will be sold.

6. Jake delivered 90 circulars in 30 minutes. At this rate, how long will it take him to deliver 135 circulars?

It will take him _____ minutes.

7. At the airport, 4 planes land every 8 minutes. At this rate, how many planes will land in 1 hour?

_____ planes will land in 1 hour.

1.
2.
3.
4.
5.
6.
7.

Perfect score: 7   My score: _____

54

Lesson 6   Percent

If n stands for a number, then $n\%$ stands for the ratio of n to 100 or $\frac{n}{100}$.

$1\% = \dfrac{1}{100}$ or $.01$ ⎪ $37\% = \dfrac{37}{100}$ or $.37$ ⎪ $125\% = \dfrac{125}{100}$ or 1.25

$5\% = $ _____ or _____ ⎪ $53\% = $ _____ or _____ ⎪ $149\% = $ _____ or _____

Complete the following.

	percent	fraction	decimal
1.	3%	_____	_____
2.	27%	_____	_____
3.	121%	_____	_____
4.	7%	_____	_____
5.	39%	_____	_____
6.	141%	_____	_____
7.	9%	_____	_____
8.	11%	_____	_____
9.	167%	_____	_____
10.	57%	_____	_____
11.	251%	_____	_____
12.	69%	_____	_____
13.	391%	_____	_____
14.	87%	_____	_____

Perfect score: 28   My score: _____

Lesson 7   Fractions and Percent

Study how to change a fraction to a percent.

$$\frac{4}{5}=\frac{n}{100} \qquad \frac{1}{8}=\frac{n}{100}$$

$$400=5n \qquad 100=8n$$

$$80=n \qquad 12\frac{1}{2}=n$$

$$\frac{4}{5}=\underline{80}\% \qquad \frac{1}{8}=\underline{12\frac{1}{2}}\%$$

Study how to change a percent to a fraction or mixed numeral.

$$175\%=\frac{175}{100}$$
$$=\frac{7}{4}\text{ or }1\frac{3}{4}$$

$$3\frac{1}{2}\%=\frac{3\frac{1}{2}}{100}$$
$$=3\frac{1}{2}\times\frac{1}{100}$$
$$=\frac{7}{2}\times\frac{1}{100}$$
$$=\frac{7}{200}$$

Complete the following.

           a   b

1. $\frac{1}{4}=$_____% $\frac{3}{8}=$_____%

2. $\frac{1}{10}=$_____% $\frac{3}{4}=$_____%

3. $\frac{1}{2}=$_____% $\frac{5}{8}=$_____%

4. $\frac{7}{10}=$_____% $\frac{2}{5}=$_____%

5. $\frac{4}{5}=$_____% $\frac{7}{8}=$_____%

Change each of the following to a fraction or mixed numeral in simplest form.

6. $10\%=$_____ $80\%=$_____

7. $160\%=$_____ $12\frac{1}{2}\%=$_____

8. $250\%=$_____ $62\frac{1}{2}\%=$_____

9. $20\%=$_____ $16\%=$_____

10. $125\%=$_____ $37\frac{1}{2}\%=$_____

Perfect score: 20   My score: _____

56

Lesson 8   Decimals and Percent

Study how to change a
decimal to a percent.

$.3 = .30 = \frac{30}{100} = \underline{\ 30\% \ }$

$1.24 = \frac{124}{100} = \underline{\ 124\% \ }$

$.375 = \frac{37.5}{100} = \underline{\ 37.5\% \ }$

$1.6 = 1.60 = \frac{160}{100} = \underline{\hspace{2cm}}\%$

$.59 = \frac{59}{100} = \underline{\hspace{2cm}}\%$

$2.125 = \frac{212.5}{100} = \underline{\hspace{2cm}}\%$

Study how to change a percent
to a decimal.

$17.6\% = \frac{17.6}{100} = \underline{\ .176 \ }$

$7.25\% = \frac{7.25}{100} = \underline{\ .0725 \ }$

$16\frac{3}{4}\% = 16.75\% = \frac{16.75}{100} = \underline{\ .1675 \ }$

$8.4\% = \frac{8.4}{100} = \underline{\hspace{2cm}}$

$9.69\% = \frac{9.69}{100} = \underline{\hspace{2cm}}$

$37\frac{1}{2}\% = 37.5\% = \frac{37.5}{100} = \underline{\hspace{2cm}}$

Complete the following.

a

	decimal	percent
1.	.2	_____
2.	1.9	_____
3.	.02	_____
4.	.36	_____
5.	1.47	_____
6.	.067	_____
7.	.123	_____
8.	1.625	_____

b

percent	decimal
52%	_____
148%	_____
5.4%	_____
8.75%	_____
183.75%	_____
$9\frac{1}{2}\%$	_____
$7\frac{1}{4}\%$	_____
$8\frac{3}{4}\%$	_____

Perfect score: 16   My score: _____

57

Lesson 9   Fractions, Decimals, and Percent

Change each fraction to a percent. Change each percent to a fraction or mixed numeral in simplest form.

	a	*b*
1.	$\frac{1}{8}=$ _____ %	$30\% =$ _____
2.	$80\% =$ _____	$\frac{1}{5}=$ _____ %
3.	$\frac{3}{5}=$ _____ %	$120\% =$ _____
4.	$87\frac{1}{2}\% =$ _____	$\frac{3}{4}=$ _____ %
5.	$\frac{1}{10}=$ _____ %	$150\% =$ _____
6.	$31\frac{1}{4}\% =$ _____	$\frac{9}{10}=$ _____ %
7.	$\frac{4}{25}=$ _____ %	$64\% =$ _____
8.	$110\% =$ _____	$\frac{9}{20}=$ _____ %

Change each decimal to a percent. Change each percent to a decimal.

9.	$.5 =$ _____ %	$4\% =$ _____
10.	$17.7\% =$ _____	$1.1 =$ _____ %
11.	$.67 =$ _____ %	$6.625\% =$ _____
12.	$8.46\% =$ _____	$1.58 =$ _____ %
13.	$.125 =$ _____ %	$4.075\% =$ _____
14.	$6.007\% =$ _____	$.312 =$ _____ %
15.	$6\frac{1}{4}\% =$ _____	$9\frac{3}{4}\% =$ _____
16.	$7\frac{3}{4}\% =$ _____	$5\frac{1}{2}\% =$ _____

Perfect score: 32   My score: _____

Lesson 10   Percent of a Number

What number is $16\frac{1}{2}\%$ of 90?	What number is 135% of 83?

$$n = 16\tfrac{1}{2}\% \times 90$$

$$= .165 \times 90$$

$$= \underline{\quad 14.85 \quad}$$

$\underline{14.85}$ is $16\frac{1}{2}\%$ of 90.

$$n = 135\% \times 83$$

$$= 1.35 \times 83$$

$$= \underline{\quad 112.05 \quad}$$

_____ is 135% of 83.

Complete the following.

a

b

1. _____ is 40% of 20.   _____ is 32% of 15.

2. _____ is 120% of 80.   _____ is 62% of 48.

3. _____ is 33% of 69.   _____ is 150% of 38.

4. _____ is $62\frac{1}{2}\%$ of 840.   _____ is 6.7% of 83.

5. _____ is 50% of $\frac{3}{8}$.   _____ is 7.8% of 65.

6. _____ is 85% of 480.   _____ is 25% of 23.6.

7. _____ is $37\frac{1}{2}\%$ of 64.   _____ is 175% of 40.

8. _____ is 6% of 112.   _____ is 9.6% of 480.

9. _____ is 80% of 540.   _____ is 12.5% of 49.8.

10. _____ is 8% of 180.   _____ is 130% of 96.

Perfect score: 20   My score: _____

Problem Solving

Solve each problem.

1. Of the building permits issued, 85% were for single-family dwellings. There were 760 permits issued. How many were for single-family dwellings?

_____ were for single-family dwellings.

2. Leona answered all the questions on a test. She had 90% of them correct. There were 40 questions in all. How many did she have correct?

She had _____ correct.

3. Of the 45 seats on the bus, 60% are filled. How many seats are filled?

_____ seats are filled.

4. An oil tank will hold 250 gallons. The tank is 80% full. How many gallons of oil are in the tank?

_____ gallons of oil are in the tank.

5. A contractor is to remove 600 cubic feet of earth. So far, 70% of the work has been done. How many cubic feet of earth have been removed?

_____ cubic feet of earth have been removed.

6. Mrs. Hughes bought a mixture of grass seed that contained 75% bluegrass seed. She purchased $2\frac{1}{2}$ pounds of grass seed in all. How many pounds of bluegrass seed did she get?

She got _____ pounds of bluegrass seed.

7. Mr. Jones' car gets 15.6 miles per gallon of gasoline. He can improve his gas mileage by 15% by getting a tune-up. By how much will his gas mileage improve with a tune-up?

His gas mileage will improve by _____ miles per gallon.

1.
2.
3.
4.
5.
6.
7.

Perfect score: 7   My score: _____

Lesson 11　Finding What Percent One Number Is of Another

25 is what percent of 40?	$\frac{3}{8}$ is what percent of $\frac{1}{2}$?

$$25 = n\% \times 40$$

$$25 = \frac{n}{100} \times 40$$

$$25 = \frac{40n}{100}$$

$$2500 = 40n$$

$$\underline{\quad 62.5 \quad} = n$$

25 is _____% of 40.

$$\frac{3}{8} = n\% \times \frac{1}{2}$$

$$\frac{3}{8} = \frac{n}{100} \times \frac{1}{2}$$

$$\frac{3}{8} = \frac{n}{200}$$

$$600 = 8n$$

$$\underline{\quad 75 \quad} = n$$

$\frac{3}{8}$ is _____% of $\frac{1}{2}$.

Complete the following.

a　　　　　　　　　　　　　　　　　b

1.　32 is _____% of　64.　　　　40 is _____% of　50.

2.　88 is _____% of　80.　　　　67 is _____% of　67.

3.　$\frac{3}{8}$ is _____% of　$\frac{3}{4}$.　　　　.8 is _____% of　3.2.

4.　$18\frac{3}{4}$ is _____% of　75.　　　　96 is _____% of　120.

5.　50 is _____% of　80.　　　　1.6 is _____% of　6.4.

6.　$\frac{2}{3}$ is _____% of　$\frac{5}{6}$.　　　　19 is _____% of　76.

7.　78 is _____% of　104.　　　　19 is _____% of　95.

8.　.72 is _____% of　.48.　　　　24 is _____% of　40.

9.　$8\frac{1}{3}$ is _____% of　$33\frac{1}{3}$.　　　　64 is _____% of　80.

10.　$6\frac{1}{4}$ is _____% of　50.　　　　.69 is _____% of　2.76.

Perfect score: 20　　My score: _____

Problem Solving

Solve each problem.

1. Last season a baseball player hit 48 home runs. So far this season he has hit 30 home runs. The number of home runs he has hit so far this season is what percent of the number of home runs he hit last season?

The number he has hit this season is _____ % of the number he hit last season.

2. In April, 175 cases of toy cars were sold. In May, 125 cases were sold. April's sales were what percent of May's sales?

April's sales were _____ % of May's sales.

3. The down payment on a bike is $15. The bike costs $75. The down payment is what percent of the cost?

The down payment is _____ % of the cost.

4. On a spelling test, Janice spelled 17 words correctly. There were 20 words on the test. What percent of the words did she spell correctly?

She spelled _____ % correctly.

5. The Andersons are planning to take a 960-mile trip. They will travel 840 miles by car. What percent of the distance will they travel by car?

They will travel _____ % of the distance by car.

6. During basketball practice, Lea attempted 30 free throws and made 21. What percent of these free-throw attempts did she make?

She made _____ % of the attempts.

7. Emily weighs 120 pounds, and Marta weighs 80 pounds. Emily's weight is what percent of Marta's weight?

Emily's weight is _____ % of Marta's weight.

1.

2.

3.

4.

5.

6.

7.

Perfect score: 7   My score: _____

Lesson 12   Finding a Number When a Percent of It Is Known

32 is 16% of what number?	1.4 is 5.6% of what number?

$$32 = 16\% \times n$$

$$32 = \frac{16}{100} \times n$$

$$32 = \frac{16\,n}{100}$$

$$3200 = 16\,n$$

$$\underline{\quad 200 \quad} = n$$

32 is 16% of _____.

$$1.4 = 5.6\% \times n$$

$$1.4 = \frac{5.6}{100} \times n$$

$$1.4 = \frac{5.6\,n}{100}$$

$$140 = 5.6\,n$$

$$\underline{\quad 25 \quad} = n$$

1.4 is 5.6% of _____.

Complete the following.

	a	b
1.	37 is 20% of _____.	92 is 50% of _____.
2.	3.4 is 25% of _____.	60 is 150% of _____.
3.	60 is 60% of _____.	9 is 30% of _____.
4.	50 is 40% of _____.	78 is 60% of _____.
5.	264 is 6% of _____.	84 is 12% of _____.
6.	18 is 75% of _____.	2.6 is 50% of _____.
7.	8.7 is 30% of _____.	72 is 80% of _____.
8.	9 is 100% of _____.	1.3 is 65% of _____.
9.	144 is 24% of _____.	2.16 is 3.6% of _____.
10.	192 is 75% of _____.	12.8 is 6.4% of _____.

Perfect score: 20   My score: _____

Problem Solving

Solve each problem.

1. Mr. Buccola has a tree which is 42 feet tall. He estimates that the tree is 75% as tall now as it will be when fully grown. How tall will the tree be when fully grown?

The tree will be _____ feet tall.

2. There are 35 boys on the school football team. This number represents 5% of the school's total enrollment. What is the school's total enrollment?

The school's total enrollment is _____.

3. Doris has read 120 pages of a library book. This is 40% of the book. How many pages are there in the book?

There are _____ pages in the book.

4. When operating at 75% capacity, a factory can produce 360 cases of nails each day. How many cases of nails can be produced each day when the factory is operating at full capacity?

_____ cases can be produced each day.

5. Martha received 212 votes for class secretary. This was 53% of the total number of votes cast. How many votes were cast?

_____ votes were cast.

6. Arlene has earned 75% of the points she needs for a prize. She has earned 660 points. How many points are needed to win a prize?

_____ points are needed.

7. Judy can throw a baseball 240 feet. This is 80% as far as her older brother can throw the ball. How far can her older brother throw the ball?

Her older brother can throw the ball _____ feet.

1.
2.
3.
4.
5.
6.
7.

Perfect score: 7   My score: _____

Lesson 13   Percent

Complete the following.

	a	*b*

1. _____ is 40% of 30.   73 is _____% of 365.

2. 26 is _____% of 50.   24 is 60% of _____.

3. 39 is 52% of _____.   _____ is 25% of 76.

4. 37 is _____% of 50.   _____ is 60% of 360.

5. 18 is 25% of _____.   69 is _____% of 276.

6. _____ is 24% of 96.   8 is 16% of _____.

7. .7 is _____% of 1.4.   _____ is 50% of 98.4.

8. 3.9 is 75% of _____.   .09 is _____% of .25.

9. _____ is 6.8% of 720.   .95 is 1.9% of _____.

10. 64 is 125% of _____.   _____ is 100% of 986.

11. 175 is _____% of 125.   98 is 150% of _____.

12. _____ is 120% of 720.   275 is _____% of 125.

13. $\frac{1}{3}$ is _____% of $\frac{5}{6}$.   30 is 75% of _____.

14. _____ is 60% of 1000.   1 is _____% of 1.

15. 15 is 50% of _____.   _____ is 75% of 2.

16. _____ is 25% of 4.   73 is 25% of _____.

Perfect score: 32   My score: _____

Problem Solving

Solve each problem.

1. There are 850 students at a school. Of these, 36% are in the eighth grade. How many students are in the eighth grade?

_____ students are in the eighth grade.

1.

2. Mail was delivered to 171 out of the 180 houses on Saylor Street. To what percent of the houses on Saylor Street was mail delivered?

Mail was delivered to _____ % of the houses.

2.

3. A savings bond costs $18.75 and can be redeemed at maturity for $25. The cost of the bond is what percent of its value at maturity?

Its cost is _____% of its value at maturity.

3.

4. Mrs. Wilson sold merchandise to 25% of the clients she contacted. She sold to 6 clients. How many clients did she contact?

She contacted _____ clients.

4.

5. A store sold 185 bicycles last month. Of those, 60% were girls' bicycles. How many girls' bicycles were sold?

_____ girls' bicycles were sold.

5.

6. Of the library books turned in today, 95% were turned in on time. There were 285 books turned in on time. What was the total number of books turned in?

There were _____ books turned in.

6.

7. The enrollment at St. Mary's School is 110% of last year's enrollment. The enrollment last year was 390. What is the enrollment this year?

The enrollment this year is _____.

7.

Perfect score: 7   My score: _____

66

CHAPTER 4 TEST

Express each of the following as a ratio in two ways as shown.

		a	b
1.	8 touchdowns in 3 games	_8 to 3_	$\frac{8}{3}$
2.	5 policemen to 4 firemen	_____	_____
3.	4 planes in 30 minutes	_____	_____
4.	5 quarts for 9 boys	_____	_____

Solve the following.

	a	b
5.	$\frac{n}{3}=\frac{12}{36}$	$\frac{4}{n}=\frac{16}{20}$
6.	$\frac{8}{9}=\frac{n}{45}$	$\frac{7}{8}=\frac{49}{n}$
7.	$\frac{18}{24}=\frac{n}{16}$	$\frac{n}{12}=\frac{4}{16}$

Complete the following.

	a	b
8.	_____ is 35% of 64.	11 is _____% of 55.
9.	18 is _____% of 25.	30 is 7.5% of _____.
10.	$1\frac{3}{4}$ is _____% of $2\frac{1}{2}$.	_____ is $12\frac{1}{2}$% of 27.
11.	_____ is 150% of 180.	2.4 is _____% of 9.6.

Perfect score: 20   My score: _____

PRE-TEST—Simple and Compound Interest

Complete the following for simple interest.

	principal	rate	time	interest
1.	$320	7%	1 year	
2.	$300	$5\frac{1}{2}$%	$\frac{1}{2}$ year	
3.	$800	10%		$80
4.	$500		$\frac{1}{4}$ year	$10
5.		16%	2 years	$192
6.	$26,000		4 years	$9360

Interest is to be compounded in each account below. Find the total amount that will be in each account after the period of time indicated.

	principal	rate	time	compounded	total amount
7.	$200	6%	2 years	annually	
8.	$100	5%	3 years	annually	
9.	$300	8%	$1\frac{1}{2}$ years	semi-annually	
10.	$400	5%	1 year	quarterly	

Perfect score: 10   My score: _____

Lesson 1   Simple Interest

If the rate of interest is 12% a year, what will the interest be on a $300 loan for $1\frac{1}{2}$ years?

$$interest = principal \times rate \times time \text{ (in years)}$$

$$
\begin{aligned}
i &= p \times r \times t \\
&= 300 \times .12 \times \frac{3}{2} \\
&= 36 \times \frac{3}{2} \\
&= 54
\end{aligned}
$$

The interest will be $_____.

If the rate of interest is $9\frac{1}{2}$% a year, what will the interest be on a $100 loan for 2 years?

$$
\begin{aligned}
i &= p \times r \times t \\
&= \underline{} \times \underline{} \times \underline{} \\
&= \underline{} \times \underline{} \\
&= \underline{}
\end{aligned}
$$

The interest will be $_____.

Find the interest for each loan described below.

	principal	rate	time	interest
1.	$250	10%	2 years	
2.	$400	12%	2 years	
3.	$550	8%	$1\frac{1}{4}$ years	
4.	$650	$11\frac{1}{2}$%	3 years	
5.	$600	16%	3 years	
6.	$500	$11\frac{1}{4}$%	1 year	
7.	$1500	15%	$1\frac{1}{3}$ years	
8.	$1000	$12\frac{1}{2}$%	3 years	
9.	$2890	14%	$2\frac{1}{2}$ years	
10.	$2600	9%	$2\frac{1}{2}$ years	

Perfect score: 10   My score: _____

Problem Solving

Solve each problem.

1. Mr. Wilkinson borrowed $600 for $1\frac{1}{2}$ years. He is to pay 9% annual interest. How much interest is he to pay?

He will pay $_____ interest.

1.

2. Dixie had $350 in a savings account for $\frac{1}{2}$ year. Interest was paid at an annual rate of 5%. How much interest did she receive?

She received $_____ interest.

2.

3. Suppose you deposit $700 in a savings account at $5\frac{1}{2}$% interest. How much interest will you receive in 1 year?

You will receive $_____.

3.

4. The Tremco Company borrowed $10,000 at 12% annual interest for a 1-year period. How much interest did the company have to pay? What was the total amount (principal + interest) the company needed to repay the loan?

The company had to pay $_____ interest.

The total amount needed was $_____.

4.

5. John borrowed $700 for 1 year. Interest on the first $300 of the loan was 18%, and interest on the remainder of the loan was 12%. How much interest did he pay?

He paid $_____ interest.

5.

6. Molly's mother borrowed $460 at 10% annual interest. What would the interest be if the loan were repaid after $\frac{1}{2}$ year? What would the interest be if the loan were repaid after $\frac{3}{4}$ year?

The interest would be $_____ for $\frac{1}{2}$ year.

The interest would be $_____ for $\frac{3}{4}$ year.

6.

Perfect score: 8   My score: _____

70

Lesson 2   Simple Interest

$36 interest paid in 2 years at a rate of 9%. Find the principal.

$$i = p \times r \times t$$
$$36 = p \times .09 \times 2$$
$$36 = .18p$$
$$\frac{36}{.18} = p$$
$$\underline{200} = p$$

The principal is $ _200_ .

$16 interest paid in 2 years on $100 principal. Find the rate.

$$i = p \times r \times t$$
$$16 = 100 \times r \times 2$$
$$16 = 200\,r$$
$$\frac{16}{200} = r$$
$$\underline{} = r$$

The rate is _____%.

$50 interest paid on $200 principal at a rate of 10%. Find the time.

$$i = p \times r \times t$$
$$50 = 200 \times .10 \times t$$
$$50 = 20t$$
$$\frac{50}{20} = t$$
$$\underline{} = t$$

The time is _____ years.

Complete the following.

	principal	rate	time	interest
1.		7%	3 years	$21
2.	$325		$1\frac{1}{2}$ years	$39
3.	$375	10%		$18.75
4.	$780	15%	2 years	
5.	$1200	9%		$216
6.	$1400		$1\frac{1}{2}$ years	$168
7.		$8\frac{1}{2}$%	$1\frac{1}{2}$ years	$446.25
8.	$8000		$2\frac{1}{2}$ years	$1500
9.	$18,050	12%		$6498
10.	$25,000	15%	3 years	

Perfect score: 10   My score: _____

Problem Solving

Solve each problem.

1. Mrs. Vernon paid $72 interest for a 2-year loan at 9% annual interest. How much money did she borrow?

She borrowed $_____.

1. _____

2. Marvin paid $63 interest for a $350 loan for $1\frac{1}{2}$ years. What was the rate of interest?

The rate of interest was _____ %.

2. _____

3. Suppose you borrow $600 at 10% interest. What period of time would you have the money if the interest is $30?

The period of time would be _____ year.

3. _____

4. Jerry had $740 in a savings account at 5% interest. The money was in the account for $\frac{1}{4}$ year. How much interest did he receive? Suppose he withdrew the principal and interest after $\frac{1}{4}$ year. How much money would he withdraw from the account?

He received $_____ interest.

He will withdraw $_____ from the account.

4. _____

5. How much must you deposit at $5\frac{1}{2}$% annual interest in order to earn $33 in 1 year?

You would need $_____ in the account.

5. _____

6. The interest on a $300 loan for 2 years is $90. What rate of interest is charged?

The rate of interest is _____%.

6. _____

7. How much must you have on deposit at $6\frac{1}{2}$% annual interest in order to earn $221 a year?

You would have to deposit $_____.

7. _____

Perfect score: 8   My score: _____

Lesson 3 Compound Interest

Interest paid on the original principal and the interest already earned is called **compound interest.**

Bev had $400 in a savings account for 3 years that paid 6% interest compounded annually. What was the total amount in her account at the end of the third year?

At the end   interest = 400 × .06 × 1 = 24.00 or $24

of 1 year: new principal = 400 + 24 = 424 or $424

At the end   interest = 424 × .06 × 1 = 25.44 or $25.44

of 2 years: new principal = 424 + 25.44 = 449.44 or $449.44

At the end   interest = 449.44 × .06 × 1 = 26.9664 or $26.97

of 3 years: total amount = ___449.44___ + ___26.97___ = ___476.41___ or $_____

Assume interest is compounded annually. Find the total amount for each of the following.

	principal	*rate*	*time*	*total amount*
1.	$500	6%	2 years	
2.	$700	$5\frac{1}{2}$%	2 years	
3.	$800	5%	3 years	
4.	$800	$6\frac{1}{2}$%	3 years	
5.	$200	9%	3 years	
6.	$1000	8%	2 years	

Perfect score: 6   My score: _____

Problem Solving

Solve each problem.

1. Trudy had $600 in a savings account for 2 years. Interest was paid at the rate of 6% compounded annually. What was the total amount in her account at the end of 2 years?

The total amount was $_____.

1.

2. Harry deposited $400 in an account that pays 5% interest compounded annually. What will be the total amount in his account after 2 years? After 3 years?

It will be $_____ after 2 years.

It will be $_____ after 3 years.

2.

3. Martha deposited $300 in an account that pays 7% interest compounded annually. Rita deposited $300 in an account at an annual rate of 7% (simple interest). After 3 years what will be the total amount in Martha's account? In Rita's account?

Martha's account will be $_____.

Rita's account will be $_____.

3.

4. Mrs. Shirley has $500 in her savings account which pays 5% interest compounded annually. What will be the value of the account after 3 years?

The value would be $_____.

4.

5. Bill deposited $300 at 6% interest compounded annually. Alice deposited $200 at $6\frac{1}{2}\%$ interest compounded annually. Who will have the greater account after 3 years? How much greater will it be?

_____ will have the greater account.

It will be $_____ greater.

5.

Perfect score: 8   My score: _____

74

Lesson 4   Compound Interest

Interest may be paid annually (each year), semi-annually (twice a year), quarterly (four times a year), monthly (every month), or daily (every day).

Ed had $100 in an account for $1\frac{1}{2}$ years that paid 6% interest compounded semi-annually. What was the total amount in his account at the end of $1\frac{1}{2}$ years?

At the end

interest = $100 \times .06 \times \frac{1}{2} = 3.00$ or $3

of $\frac{1}{2}$ year: new principal = $100 + 3 = 103$ or $103

At the end

interest = $103 \times .06 \times \frac{1}{2} = 3.09$ or $3.09

of 1 year: new principal = $103 + 3.09 = 106.09$ or $106.09

At the end

interest = $106.09 \times .06 \times \frac{1}{2} = 3.1827$ or $3.18

of $1\frac{1}{2}$ years: total amount = _____ + _____ = _____ or $_____

Find the total amount for each of the following.

	principal	rate	time	compounded	total amount
1.	$200	6%	$1\frac{1}{2}$ years	semi-annually	
2.	$300	5%	2 years	semi-annually	
3.	$100	5%	1 year	quarterly	
4.	$400	7%	$\frac{3}{4}$ year	quarterly	
5.	$500	8%	4 months	monthly	
6.	$600	9%	$\frac{1}{4}$ year	monthly	

Perfect score: 6   My score: _____

Problem Solving

Solve each problem.

1. Mrs. Fauler had $600 in a savings account that paid 5% interest compounded semi-annually. What was the value of her account after $1\frac{1}{2}$ years?

The value was $_____.

2. How much interest would $3000 earn in 2 years at 7% interest compounded semi-annually?

It would earn $_____ interest.

3. Suppose $100 were deposited in each savings account with rates of interest as follows:
   Account **A**—6% compounded annually
   Account **B**—6% compounded semi-annually
   Account **C**—6% compounded quarterly
What would be the value of each account after 1 year?

$_____ will be in account **A**.

$_____ will be in account **B**.

$_____ will be in account **C**.

4. Assume $200 was deposited in a 2-year account at 9%. How much more interest would be in the account if the interest were compounded annually rather than computed as simple interest?

There would be $_____ more in the account.

5. Account **A** has $500 at 8% interest compounded quarterly. Account **B** has $500 at 8% interest compounded semi-annually. Which account will have a greater amount of money after 1 year? How much more?

Account _____ will have more money.

It will have $_____ more.

1.
2.
3.
4.
5.

Perfect score: 8   My score: _____

76

CHAPTER 5 TEST

Complete the following for simple interest.

	principal	rate	time	interest
1.	$150	15%	3 years	
2.	$700	$8\frac{1}{2}$%	2 years	
3.	$645		$\frac{1}{4}$ year	$19.35
4.	$540	10%		$135
5.		$9\frac{1}{2}$%	2 years	$729.60
6.	$1800		2 years	$540

Interest is to be compounded in each account below. Find the total amount that will be in each account after the period of time indicated.

	principal	rate	time	compounded	total amount
7.	$300	7%	2 years	annually	
8.	$600	5%	3 years	annually	
9.	$500	6%	2 years	semi-annually	
10.	$400	9%	$\frac{1}{4}$ year	monthly	

Perfect score: 10   My score: _____

Circle the unit you would use to measure each of the following.

1. capacity of a tank   *meter*   *liter*   *gram*

2. length of a string   *centimeter*   *centiliter*   *centigram*

3. weight of an ant   *millimeter*   *milliliter*   *milligram*

Write *1000, .01,* or *.001* to make each sentence true.

4. The prefix *milli* means _____.

5. The prefix *kilo* means _____.

6. The prefix *centi* means _____.

Measure each line segment to the nearest unit indicated.

7. _____ centimeters   ████████████████

8. _____ millimeters   ██████████████████

Complete the following.

	a	*b*
9.	1 meter = _____ kilometer	100 millimeters = _____ centimeters
10.	2 liters = _____ milliliters	2 kilograms = _____ grams
11.	.5 gram = _____ milligrams	300 centimeters = _____ meters
12.	1.4 kiloliters = _____ liters	.05 kilometer = _____ meters

Perfect score: 16   My score: _____

Lesson 1   Metric Measurement

A **meter** is a unit of *length*.

A **liter** is a unit of *capacity*.

A **gram** is a unit of *mass* (or *weight*).

Kilo means **1000**.	*Kilometer* means ___1000___ meters.
Hecto means **100**.	*Hectoliter* means ___100___ liters.
Deka means **10**.	*Dekagram* means ___10___ grams.
Deci means **.1**.	*Decimeter* means _____ meter.
Centi means **.01**.	*Centiliter* means _____ liter.
Milli means **.001**.	*Milligram* means _____ gram.

The most commonly used prefixes are *kilo, centi,* and *milli.*

Tell whether the following would be measured in *meters, liters,* or *grams.*

	a	b
1.	amount of juice in a glass _____	weight of a pencil _____
2.	distance a baseball is thrown _____	length of a bus _____
3.	amount of fuel in a truck _____	weight of a bird _____

Complete the following as shown.

4. Kiloliter means ___1000 liters___.   Kilogram means _____.

5. Centigram means _____.   Centimeter means _____.

6. Milliliter means _____.   Millimeter means _____.

Name two things that could be measured with each of the following.

7. meters   _____   _____

8. liters   _____   _____

9. grams   _____   _____

Perfect score: 17   My score: _____

79

Lesson 2 Length

A **meter** (m) is approximately 39.37 inches long.

To name a unit of length other than the meter, a *prefix* is attached to the word meter. This prefix denotes the relationship of that particular unit to the meter.

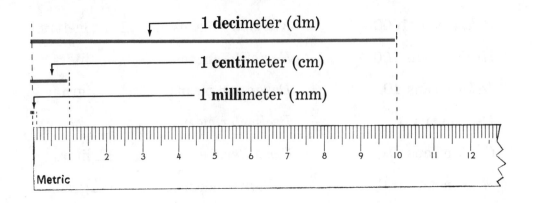

1 mm = __.001__ m   1 dm = _____ m

1 cm = _____ m   1 **kilo**meter (km) = 1000 m

In each pair of measurements below, draw a ring around the measurement for the greater length.

	a	*b*	*c*
1.	1 km ; 1 dm	1 dm ; 1 cm	1 km ; 1 mm
2.	1 mm ; 1 dm	1 cm ; 1 km	1 cm ; 1 mm

Complete the following.

	a	*b*	*c*
3.	1 m = _____ cm	1 m = _____ mm	1 m = _____ dm
4.	.01 m = _____ cm	.001 m = _____ mm	.1 m = _____ dm
5.	1000 m = _____ km	1 m = _____ km	.001 km = _____ m

Perfect score: 15   My score: _____

Lesson 3   Units of Length

To change from	to millimeters, multiply by	to centimeters, multiply by	to meters, multiply by	to kilometers, multiply by
millimeters		.1	.001	.000001
centimeters	10		.01	.00001
meters	1000	100		.001
kilometers	1,000,000	100,000	1000	

Using the table makes it easy to complete exercises like the following.

8.43 km = ____?____ m

1 km = 1000 m

8.43 km = (8.43 × 1000) m

8.43 km = ___8430___ m

75 mm = ____?____ cm

1 mm = .1 cm

75 mm = (75 × .1) cm

75 mm = _____ cm

Complete.

	a	b
1.	5 km = _____ m	.452 km = _____ m
2.	38 m = _____ km	948 m = _____ km
3.	7.5 m = _____ cm	80 m = _____ cm
4.	4 cm = _____ m	75 cm = _____ m
5.	92 cm = _____ mm	4.86 cm = _____ mm
6.	92 mm = _____ cm	7 mm = _____ cm
7.	.5 m = _____ mm	.003 m = _____ mm
8.	92 mm = _____ m	3600 mm = _____ m

9. A city block is about 200 meters long. How long is a city block in kilometers?   _____

10. How long would five city blocks be in meters? _____   In kilometers? _____

Perfect score: 19   My score: _____

Lesson 4   Capacity

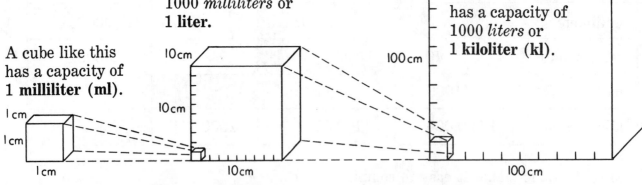

A cube like this has a capacity of **1 milliliter (ml)**.

A cube like this has a capacity of 1000 *milliliters* or **1 liter.**

A cube like this has a capacity of 1000 *liters* or **1 kiloliter (kl).**

| 1000 ml = 1 liter |
| .001 liter = 1 ml |

| 1000 liters = 1 kl |
| .001 kl = 1 liter |

Underline the measurement for the greater amount.

	a	*b*
1.	10 liters, 10 kiloliters	100 milliliters, 1 kiloliter
2.	.1 kiloliter, 1000 liters	10 milliliters, 1 liter
3.	1000 liters, 10,000 milliliters	.001 kiloliter, 1 milliliter
4.	500 liters, 1 kiloliter	700 milliliters, 1 liter

Complete the following.

5. 1 liter = _____ milliliters   .1 liter = _____ milliliters

6. 1 kiloliter = _____ liters   .01 kiloliter = _____ liters

7. .001 kiloliter = _____ liter   1000 milliliters = _____ liter

8. 100 liters = _____ kiloliter   10 kiloliters = _____ liters

Perfect score: 16   My score: _____

82

Lesson 5   Units of Capacity

1.2 kl = _____?_____ liters

   1 kl = 1000 liters

   1.2 kl = (1.2 × 1000) liters

   1.2 kl = __1200__ liters

54 liters = _____?_____ kl

   1 liter = .001 kl

   54 liters = (54 × .001) kl

   54 liters = _____ kl

Complete the following.

	a	*b*
1.	6.4 liters = _____ ml	6000 ml = _____ liters
2.	25 kl = _____ liters	752 liters = _____ kl
3.	78 liters = _____ ml	529 ml = _____ liter
4.	.986 kl = _____ liters	42 liters = _____ kl
5.	7.5 liters = _____ ml	7.5 ml = _____ liter
6.	7.5 kl = _____ liters	7.5 liters = _____ kl

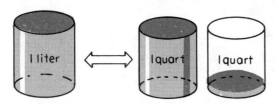

A liter is slightly more than a quart.

A teaspoon has a capacity of about 5 milliliters.

7. Is 1 quart more than or less than 1 liter? _____

8. A recipe calls for 2 teaspoons of vanilla. How many milliliters would that be?

It would be _____ milliliters.

8.

9. Three teaspoons equal 1 tablespoon. How many milliliters are in 1 tablespoon? What part of a liter is 1 tablespoon?

_____ milliliters are in 1 tablespoon.

1 tablespoon is _____ liter.

9.

Perfect score: 16   My score: _____

83

Lesson 6   Units of Mass

An aspirin tablet weighs about 350 **milligrams** (mg).

1 milliliter of water weighs 1 **gram** (g).

1 liter of water weighs 1 **kilogram** (kg).

1000 mg = 1 g
.001 g = 1 mg

1000 g = 1 kg
.001 kg = 1 g

65 g = ___?___ mg
1 g = 1000 mg
65 g = (65 × 1000) mg
65 g = _65,000_ mg

250 g = ___?___ kg
1 g = .001 kg
250 g = (250 × .001) kg
250 g = _____ kg

Complete the following.

	a	b
1.	26 g = _____ mg	6.2 g = _____ mg
2.	75.2 mg = _____ g	2420 mg = _____ g
3.	89 kg = _____ g	7.5 kg = _____ g
4.	835 g = _____ kg	5.6 g = _____ kg
5.	60.5 g = _____ mg	60.5 g = _____ kg

6. A teaspoon holds about 5 milliliters of water. How much would the 5 milliliters of water weigh in grams? In milligrams?

It would weigh _____ grams.

It would weigh _____ milligrams.

6.

7. A nickel weighs about 5 grams. What is the weight of 200 nickels in grams? In kilograms?

200 nickels weigh about _____ grams.

200 nickels weigh about _____ kilogram.

7.

Perfect score: 14   My score: _____

84

Lesson 7   Problem Solving

Solve each problem.

1. A pitcher contained 1.2 liters of milk. You used 250 milliliters of milk from the pitcher. How many milliliters of milk are left in the pitcher?

_____ milliliters are left.

2. Simon says he is 1.6 meters tall. Rosa says she is 162 centimeters tall. Who is taller? How many centimeters taller?

_____ is _____ centimeters taller.

3. A jet flew 1 kilometer on 15 liters of fuel. How many kiloliters of fuel are needed for the jet to fly 3000 kilometers?

The jet would need _____ kiloliters of fuel.

4. Two liters of grape juice will fill 8 glasses of the same size. What is the capacity of each glass in milliliters?

Each glass has a capacity of _____ milliliters.

5. Tim bought 6 kilograms of meat for $25.20. What was the cost per kilogram?

The cost was $_____ per kilogram.

6. During a contest, frog A jumped 59.3 centimeters. Frog B jumped 590 millimeters. Which frog jumped farther? How many centimeters farther?

Frog _____ jumped _____ centimeter farther.

7. Al drove 158 kilometers. Bea drove 230 kilometers. How much farther than Al did Bea drive?

She drove _____ kilometers farther.

1.

2.

3.

4.

5.

6.

7.

Perfect score: 9   My score: _____

85

Lesson 8   Temperature

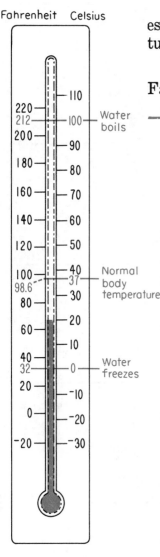

Fahrenheit   Celsius

By placing the edge of a paper across the scales, you can estimate equivalent temperatures on the scales. The temperature outside is 20° C. What is the temperature in degrees

Fahrenheit? ___68___ ° F

_____

Complete each of the following.

1. At what Fahrenheit temperature does water freeze? _____° F

2. At what Celsius temperature does water freeze? _____° C

3. At what Fahrenheit temperature does water boil? _____° F

4. At what Celsius temperature does water boil? _____° C

5. What is normal body temperature in degrees Celsius? _____° C

6. What is normal body temperature in degrees Fahrenheit?   _____° F

Circle the correct answer.

7.	swimming weather	28° Fahrenheit	28° Celsius
8.	snow-skiing weather	10° Celsius	10° Fahrenheit
9.	water-skiing weather	86° Fahrenheit	86° Celsius
10.	shirt-sleeve weather	30° Fahrenheit	30° Celsius
11.	water would be frozen	28° Celsius	28° Fahrenheit
12.	water would be boiling	112° Celsius	112° Fahrenheit

What outdoor activity might be appropriate for each temperature given below?

13. 25° Celsius _____

14. 25° Fahrenheit _____

Perfect score: 14   My score: _____

CHAPTER 6 TEST

Measure each line segment to the nearest unit as indicated.

1. _____ centimeters ▬▬▬▬▬▬▬▬

2. _____ millimeters ▬▬▬▬▬

Complete the following.

| | |
| a | b |

3. 25 cm = _____ mm   35 m = _____ cm

4. 6 km = _____ m   7 m = _____ km

5. 260 cm = _____ m   600 mm = _____ m

6. 7.5 m = _____ mm   2.5 mm = _____ cm

7. 12 liters = _____ ml   13.5 ml = _____ liter

8. 5.4 kl = _____ liters   1200 liters = _____ kl

9. .045 liter = _____ ml   260 liters = _____ kl

10. 58 kg = _____ g   400 g = _____ kg

11. 3000 mg = _____ g   3.8 kg = _____ g

12. .6 g = _____ mg   50 mg = _____ g

13. Water freezes at _____° Celsius or at _____° Fahrenheit.

14. Water boils at _____° Celsius or at _____° Fahrenheit.

Solve each problem.

15. 5.5 kiloliters of water is in a tank. If 3200 liters of water is used, how many liters will be in the tank? How many kiloliters is that?

There will be _____ liters in the tank.

That is _____ kiloliters.

16. Inez jumped 1.45 meters. Ed jumped 138 centimeters. Who jumped farther? How much farther?

_____ jumped _____ centimeters farther.

15.

16.

Perfect score: 30   My score: _____

PRE-TEST—Measurement and Approximation

Complete the following.

	a	*b*
1.	48 in.=_____ ft	4 ft 8 in. =_____ in.
2.	48 hours =_____ days	5 yd 1 ft =_____ ft
3.	3 hours =_____ min	5 lb 3 oz =_____ oz
4.	8 qt =_____ gal	5 min 4 sec =_____ sec
5.	64 oz =_____ lb	3 gal 2 qt =_____ qt
6.	16 cups =_____ pt	7 qt 1 pt =_____ pt

Add, subtract, or multiply.

	a	*b*	*c*
7.	8 ft 3 in. +2 ft 4 in.	3 min 49 sec −1 min 27 sec	2 lb 3 oz ×4
8.	5 yd 2 ft +2 yd 2 ft	5 gal 2 qt −2 gal 3 qt	1 lb 4 oz ×7

Round as indicated.

	a *nearest ten*	*b* *nearest hundred*	*c* *nearest thousand*
9. 8324	_____	_____	_____
10. 74485	_____	_____	_____

Perfect score: 24   My score: _____

Lesson 1 Linear Measurement

| 12 in.=1 ft |
| 3 ft =1 yd |
| 36 in.=1 yd |

$6\frac{1}{2}$ ft = ___?___ in.

1 ft = 12 in.
$6\frac{1}{2}$ ft = $(6\frac{1}{2} \times 12)$ in.

$6\frac{1}{2}$ ft = ___78___ in.

48 in.= ___?___ ft

12 in. = 1 ft
48 in. = (48 ÷ 12) ft

48 in. = _____ ft

4 yd 2 ft = ___?___ ft

Since 1 yd = 3 ft
4 yd = 4 × 3 or 12 ft
4 yd 2 ft = (12 + 2) ft

4 yd 2 ft = _____ ft

Complete the following.

	a	b
1.	48 in. = _____ ft	5 ft 4 in. = _____ in.
2.	7 ft = _____ in.	6 yd 1 ft = _____ ft
3.	42 ft = _____ yd	2 yd 9 in. = _____ in.
4.	5 yd = _____ ft	12 ft 9 in. = _____ in.
5.	72 in. = _____ yd	5 yd 2 ft = _____ ft
6.	7 yd = _____ in.	4 ft 7 in. = _____ in.
7.	108 in. = _____ ft	6 yd 5 in. = _____ in.
8.	126 in. = _____ yd	3 yd 2 ft = _____ ft
9.	$3\frac{3}{4}$ ft = _____ in.	8 ft $9\frac{1}{2}$ in. = _____ in.
10.	27 in. = _____ ft	2 ft $3\frac{1}{2}$ in.= _____ in.
11.	$5\frac{1}{2}$ yd = _____ in.	1 yd $5\frac{1}{4}$ in.= _____ in.

Perfect score: 22   My score: _____

89

Problem Solving

Solve each problem.

1. The Higgins' fence is 60 inches high. What is the height of the fence in feet?

The height of the fence is _____ feet.

2. Mr. Baxter is 6 feet 2 inches tall. How many inches tall is he?

He is _____ inches tall.

3. Grace purchased 2 yards of ribbon. How many inches of ribbon did she purchase?

She purchased _____ inches of ribbon.

4. A football field is 53 yards 1 foot wide. What is the width of the field in feet?

The width of the field is _____ feet.

5. Hank threw a baseball 180 feet. How many yards did he throw the baseball?

He threw the baseball _____ yards.

6. Mr. Kelly's lot is 20 yards wide. What is the width of the lot in feet?

The width of the lot is _____ feet.

7. Harold has a piece of wire 108 inches long. What is the length of the wire in yards?

The length of the wire is _____ yards.

8. When the high-jump bar is set at 5 feet 6 inches, what is the height of the bar in inches?

The height is _____ inches.

9. Mrs. Avilla's garage door is 7 feet wide. What is the width of the door in inches?

The width of the door is _____ inches.

1.

2.

3.

4.

5.

6.

7.

8.

9.

Perfect score: 9   My score: _____

Lesson 2   Liquid Measure, Time, Weight

2 cups = 1 pt
2 pt = 1 qt
4 qt = 1 gal

60 sec = 1 min
60 min = 1 hour
24 hours = 1 day

16 oz = 1 lb
2000 lb = 1 T

300 sec = __?__ min

1 min = 60 sec
300 sec = (300 ÷ 60) min

300 sec = _____ min

5 lb 14 oz = __?__ oz

1 lb = 16 oz
5 lb = 5 × 16 or 80 oz
5 lb 14 oz = (80 + 14) oz

5 lb 14 oz = _____ oz

Complete the following.

	a	*b*
1.	5½ min = _____ sec	2 min 14 sec = _____ sec
2.	48 oz = _____ lb	3 hours 40 min = _____ min
3.	15 qt = _____ gal	5 gal 3 qt = _____ qt
4.	3½ hours = _____ min	3 qt 1 pt = _____ pt
5.	3 T = _____ lb	2 pt 1 cup = _____ cups
6.	6 pt = _____ qt	6 lb 5 oz = _____ oz
7.	72 min = _____ hours	3 min 51 sec = _____ sec
8.	4 pt = _____ cups	10 lb 6 oz = _____ oz
9.	96 hours = _____ days	5 hours 16 min = _____ min
10.	4 lb = _____ oz	2 hours 15 min = _____ min
11.	3 days = _____ hours	3 gal 2 qt = _____ qt

Perfect score: 22   My score: _____

Problem Solving

Solve each problem.

1. It took Jeromy 3 minutes 25 seconds to run around the block. How many seconds did it take him to run around the block?

It took him _____ seconds.

2. The capacity of a container is 2 gallons 1 quart. What is the capacity of the container in quarts?

The capacity is _____ quarts.

3. Last month 64 quarts of milk were delivered to the Collins' house. How many gallons of milk was that?

That was _____ gallons of milk.

4. Mrs. Johnson purchased a 2-pound 8-ounce can of shortening. How many ounces of shortening did she purchase?

She purchased _____ ounces of shortening.

5. Mr. Singer showed his class a film which lasted 120 minutes. How many hours did the film last?

The film lasted _____ hours.

6. The cooling system of Mr. Bigg's car has a capacity of 8 quarts 1 pint. What is the capacity of the cooling system in pints?

The capacity is _____ pints.

7. The Smiths' baby weighed 6 pounds 7 ounces at birth. What was the weight of the baby in ounces?

The baby's weight was _____ ounces.

8. An elephant at the zoo weighs 2 tons. What is the weight of the elephant in pounds?

The elephant weighs _____ pounds.

1.

2.

3.

4.

5.

6.

7.

8.

Perfect score: 8   My score: _____

92

Lesson 3   Adding Measures

$$\begin{array}{r} 3 \text{ ft } 9 \text{ in.} \\ +5 \text{ ft } 7 \text{ in.} \\ \hline 16 \text{ in.} \end{array}$$

$(9+7)$ in. = _____ in.

$$\begin{array}{r} 3 \text{ ft } 9 \text{ in.} \\ +5 \text{ ft } 7 \text{ in.} \\ \hline 16 \text{ in.} \end{array}$$

16 in. = $(12+4)$ in.

= 1 ft 4 in.

$$\begin{array}{r} \overset{1}{3} \text{ ft } 9 \text{ in.} \\ +5 \text{ ft } 7 \text{ in.} \\ \hline 9 \text{ ft } 16 \text{ in.} \\ 4 \end{array}$$

$(1+3+5)$ ft = _____ ft

Complete the following.

	a	*b*
1.	15 in. = 1 ft _____ in.	73 sec = 1 min _____ sec
2.	3 pt = 1 qt _____ pt	27 oz = 1 lb _____ oz
3.	79 min = 1 hour _____ min	5 ft = 1 yd _____ ft
4.	5 qt = 1 gal _____ qt	3 cups = 1 pt _____ cup

Find each sum.

	a	*b*	*c*	*d*
5.	7 yd 1 ft +2 yd 1 ft	3 min 14 sec +2 min 29 sec	3 gal 1 qt +2 gal 2 qt	7 ft 4 in. +2 ft 3 in.
6.	3 lb 4 oz +2 lb 7 oz	8 ft 7 in. +2 ft 3 in.	3 hours 21 min +2 hours 16 min	5 pt 1 cup +3 pt 1 cup
7.	3 yd 2 ft +2 yd 2 ft	7 lb 9 oz +5 lb 13 oz	9 min 7 sec +3 min 6 sec	5 hours 32 min +2 hours 45 min
8.	3 lb 9 oz +6 lb 8 oz	5 gal 3 qt +1 gal 2 qt	1 min 17 sec +4 min 53 sec	3 lb 8 oz +2 lb 12 oz
9.	9 ft 7 in. +2 ft 5 in.	7 lb 3 oz +1 lb 13 oz	3 pt 1 cup +2 pt 1 cup	9 yd 1 ft +3 yd 2 ft

Perfect score: 28   My score: _____

Problem Solving

Solve each problem.

1. The first feature at the Rex Theater lasts 1 hour 45 minutes. The second lasts 1 hour 36 minutes. The features are shown one after the other. How long will the double feature last?

It will last _____ hours _____ minutes.

2. Jerry has a board 3 feet 9 inches long and Lee has a board 6 feet 5 inches long. What will be the total length of the boards if they are placed end-to-end?

The total length will be _____ feet _____ inches.

3. At birth, one of the Jones twins weighed 6 pounds 5 ounces. The other weighed 5 pounds 14 ounces. What was the total weight of the twins?

The total weight was _____ pounds _____ ounces.

4. To make a cleaning solution, 1 gallon 3 quarts of concentrate were mixed with 4 gallons 3 quarts of water. What amount of cleaning solution was made?

_____ gallons _____ quarts of cleaning solution were made.

5. Two pipes are joined together. One pipe is 6 feet 7 inches long and the other 4 feet 10 inches long. What is the total length of the pipes?

The total length is _____ feet _____ inches.

6. Mrs. Little purchased a 3-pound 8-ounce roast and 2 pounds 13 ounces of hamburger. How much meat did she purchase altogether?

She purchased _____ pounds _____ ounces of meat.

7. Jim's sister ran the first mile of a 2-mile race in 4 minutes 35 seconds. She ran the second mile in 4 minutes 53 seconds. What was her time for the race?

Her time was _____ minutes _____ seconds.

| 1. |
| 2. |
| 3. |
| 4. |
| 5. |
| 6. |
| 7. |

Perfect score: 7   My score: _____

94

Lesson 4   Subtracting Measures

$$\begin{array}{r} \overset{6}{7}\text{ ft } \overset{16}{4}\text{ in.} \\ -2\text{ ft } 7\text{ in.} \end{array}$$

7 ft 4 in. = (6+1) ft + 4 in.
= 6 ft (12+4) in.
= 6 ft 16 in.

$$\begin{array}{r} \overset{6}{7}\text{ ft } \overset{16}{4}\text{ in.} \\ -2\text{ ft } 7\text{ in.} \\ \hline 9\text{ in.} \end{array}$$

(16−7) in. = _____ in.

$$\begin{array}{r} \overset{6}{7}\text{ ft } \overset{16}{4}\text{ in.} \\ -2\text{ ft } 7\text{ in.} \\ \hline 4\text{ ft } 9\text{ in.} \end{array}$$

(6−2) ft = _____ ft

Complete the following.

	a	b
1.	15 ft 4 in. = 14 ft _____ in.	4 lb 3 oz = 3 lb _____ oz
2.	7 gal 2 qt = 6 gal _____ qt	6 yd 2 ft = 5 yd _____ ft
3.	5 qt 1 pt = 4 qt _____ pt	2 min 45 sec = 1 min _____ sec
4.	4 hours 3 min = 3 hours _____ min	3 pt 1 cup = 2 pt _____ cups

Find each difference.

	a	b	c	d
5.	9 ft 7 in. −3 ft 6 in.	7 lb 14 oz −3 lb 9 oz	3 hours 45 min −2 hours 19 min	5 gal 3 qt −1 gal 2 qt
6.	16 ft 9 in. −9 ft 9 in.	8 yd 2 ft −3 yd 1 ft	6 qt 1 pt −3 qt 1 pt	8 min 27 sec −5 min 16 sec
7.	15 lb 7 oz −12 lb 9 oz	2 pt 1 cup −1 pt 1 cup	5 hours 19 min −3 hours 45 min	8 ft 4 in. −2 ft 9 in.
8.	3 gal 1 qt −1 gal 2 qt	8 yd 1 ft −5 yd 2 ft	16 lb 5 oz −8 lb 11 oz	9 min 6 sec −2 min 8 sec
9.	6 min 0 sec −2 min 51 sec	8 gal −4 gal 1 qt	8 lb −4 lb 11 oz	7 pt −2 pt 1 cup

Perfect score: 28   My score: _____

Problem Solving

Solve each problem.

1. Mrs. Winters purchased a 7-pound 8-ounce ham and a 16-pound 4-ounce turkey. How much more did the turkey weigh than the ham?

It weighed _____ pounds _____ ounces more.

2. The first game of a doubleheader lasted 2 hours 48 minutes. The second game lasted 3 hours 19 minutes. How much longer did the second game last than the first?

The second game lasted _____ minutes longer.

3. Claudia high jumped 4 feet 9 inches. Renee high jumped 5 feet 2 inches. How much higher did Renee jump?

Renee jumped _____ inches higher.

4. A service station had 9 gallons of motor oil. In one day 5 gallons 3 quarts was sold. How much motor oil was left?

_____ gallons _____ quart of oil was left.

5. There are 4 gallons 2 quarts of paint in a bucket. Suppose 1 gallon 3 quarts were used. How much paint would be left in the bucket?

_____ gallons _____ quarts of paint will be left.

6. The Ace Factory operates 3 hours 30 minutes in the morning and 4 hours 15 minutes in the afternoon. How much longer does the factory operate in the afternoon than in the morning?

It operates _____ minutes longer in the afternoon.

7. Julie is 4 feet 7 inches tall and her boyfriend is 5 feet 9 inches tall. How much shorter is Julie than her boyfriend?

She is _____ foot _____ inches shorter.

1.	
2.	
3.	
4.	
5.	
6.	
7.	

Perfect score: 7   My score: _____

96

Lesson 5   Multiplying Measures

NAME _____

$$\begin{array}{r} 3\text{ ft }5\text{ in.}\\ \times 4\\ \hline 20\text{ in.}\end{array}$$

(4×5) in. = _____ in.

$$\begin{array}{r} 3\text{ ft }5\text{ in.}\\ \times 4\\ \hline 20\text{ in.}\end{array}$$

20 in. = _____ ft _____ in.

$$\begin{array}{r} 3\text{ ft }5\text{ in.}\\ \times 4\\ \hline 13\text{ ft }20\text{ in.}\end{array}$$

$[(4\times3)+1]$ ft = _____ ft

Find each product.

	a	b	c
1.	4 ft 2 in. ×4	3 min 12 sec ×4	3 gal 1 qt ×3
2.	3 hours 20 min ×3	5 lb 2 oz ×5	7 yd 1 ft ×2
3.	5 ft 2 in. ×9	6 gal 2 qt ×7	5 lb 4 oz ×6
4.	7 lb 5 oz ×9	6 ft 7 in. ×6	9 min 36 sec ×5
5.	3 yd 2 ft ×8	5 min 20 sec ×6	3 lb 9 oz ×3
6.	2 lb 3 oz ×5	1 hour 10 min ×6	6 qt 1 pt ×3
7.	5 yd 1 ft ×8	4 ft 8 in. ×6	7 lb 3 oz ×9

Perfect score: 21   My score: _____

97

Problem Solving

Solve each problem.

1. Six tables are to be placed end to end. Each table is 4 feet 3 inches long. What will be the total length of the tables?

The total length will be _____ feet _____ inches.

2. It takes 2 minutes 15 seconds to assemble a doo-dad. How long will it take to assemble 3 doodads?

It will take _____ minutes _____ seconds.

3. Each doodad weighs 3 pounds 9 ounces. There are 6 doodads in each case. How much will the doodads in a case weigh?

They will weigh _____ pounds _____ ounces.

4. A laundry purchased 5 large bottles of detergent. Each bottle contained 2 quarts 1 pint. How much detergent was purchased?

They purchased _____ quarts _____ pint of detergent.

5. Mr. Mitchell purchased 8 boards. Each board was 8 feet 4 inches long. What was the total length of the boards he purchased?

The total length was _____ feet _____ inches.

6. Nine cases of art supplies are to be shipped. Each case weighs 9 pounds 8 ounces. How much will the shipment weigh?

It will weigh _____ pounds _____ ounces.

7. Each shift at the Kempf Factory lasts 7 hours 30 minutes. There are 9 shifts each week. How long does the factory operate each week?

The factory operates _____ hours _____ minutes each week.

1.	
2.	
3.	
4.	
5.	
6.	
7.	

Perfect score: 7   My score: _____

98

Lesson 6   Measurement

Complete the following.

	a	*b*
1.	2 hours = _____ min	3 ft 7 in. = _____ in.
2.	5 lb = _____ oz	2 lb 11½ oz = _____ oz
3.	7 qt = _____ gal	3 min 38 sec = _____ sec
4.	72 in = _____ ft	4 qt 1 pt = _____ pt
5.	240 sec = _____ min	5 yd 2 ft = _____ ft

Add, subtract, or multiply.

	a	*b*	*c*
6.	3 ft 7 in. +5 ft 4 in.	4 lb 14 oz −2 lb 7 oz	7 qt 1 pt ×3
7.	16 hours 25 min +4 hours 35 min	6 ft 4 in. ×5	6 gal −2 gal 3 qt

Solve each of the following.

8. A truck is 8 feet 6 inches high. The bottom of an overpass is 11 feet 3 inches above the road. How much clearance will be between the top of the truck and the overpass?

8.

There will be _____ feet _____ inches clearance.

9. It is 366 feet from home plate to the right-field fence at the foul pole. What is this distance in yards?

9.

It is _____ yards.

Perfect score: 18   My score: _____

Lesson 7   Rounding Numbers

Round 32 to the nearest ten.

    32 is nearer 30 than 40.
32 rounded to the nearest

ten is ____30____.

Round 75 to the nearest ten.

    75 is as near 70 as 80.
In such cases, use the *greater multiple* of ten.
75 rounded to the nearest

ten is ____80____.

Round 4769 to the nearest hundred.

    4769 is nearer 4800 than 4700.
4769 rounded to the nearest hundred

is _____.

Round 4500 to the nearest thousand.

    4500 is as near 4000 as 5000.
In such cases, use the *greater multiple* of one thousand.
4500 rounded to the nearest

thousand is _____.

Round to the nearest ten.

	a	b	c
1.	28 _____	73 _____	85 _____
2.	244 _____	477 _____	655 _____
3.	1696 _____	2792 _____	8245 _____

Round to the nearest hundred.

	a	b	c
4.	321 _____	479 _____	550 _____
5.	1459 _____	2628 _____	1650 _____
6.	24136 _____	35282 _____	47350 _____

Round to the nearest thousand.

	a	b	c
7.	4325 _____	6782 _____	7500 _____
8.	5943 _____	8399 _____	8500 _____
9.	16482 _____	27501 _____	43500 _____

Perfect score: 27   My score: _____

Lesson 8   Estimating Sums and Differences

Estimate the sum of 744 and 378.	*estimated sum*	*actual sum*
744 —— to the nearest hundred ——→	700	744
+378 —— to the nearest hundred ——→	+400	+378
	1100	1122

To estimate the sum of 6375 and 8678, round 6375 to _____ and 8678 to _____.

The estimated sum would be 6000 + 9000 or _____.

Estimate the difference between 6232 and 2948.	*estimated difference*	*actual difference*
6232 —— to the nearest thousand ——→	6000	6232
−2948 —— to the nearest thousand ——	−3000	−2948
	3000	3284

To estimate the difference between 38735 and 12675, round 38735 to _____ and 12675

to _____. The estimated difference would be 40000 − 10000 or _____.

Estimate each sum or difference. Then find each sum or difference.

	a	*estimate*	*b*	*estimate*	*c*	*estimate*
1.	739 +435	_____	678 −245	_____	743 +825	_____
2.	7254 −1326	_____	1375 +6427	_____	2795 −1246	_____
3.	7524 +3542	_____	6852 −4526	_____	7689 +3824	_____
4.	25243 −12675	_____	76425 +23142	_____	95245 −58624	_____

Perfect score: 24   My score: _____

Lesson 9 Estimating Products

Study how to estimate the product of 187 and 63.

		estimated product	actual product
187	—— to the nearest hundred ——→	200	187
×63	—— to the nearest ten ——→	×60	×63
		12000	561
			11220
			11781

To estimate 86 × 224, round 86 to _____ and 224 to _____.

The estimated product would be 90 × 200 or _____.

Write the estimated product on each ___. Then find each product.

	a	*b*	*c*
1.	72 ×38 _____	91 ×57 _____	55 ×65 _____
2.	69 ×48 _____	56 ×78 _____	75 ×66 _____
3.	84 ×63 _____	93 ×43 _____	74 ×45 _____
4.	125 ×78 _____	469 ×36 _____	724 ×63 _____
5.	427 ×43 _____	825 ×73 _____	974 ×47 _____

Perfect score: 30   My score: _____

CHAPTER 7 TEST

NAME _____

Complete the following.

	a	*b*
1.	3 ft = _____ in.	3 yd 2 ft = _____ ft
2.	90 min = _____ hours	4 lb 11 oz = _____ oz
3.	5 lb = _____ oz	2 hours 27 min = _____ min
4.	96 hours = _____ days	3 ft 9 in. = _____ in.
5.	3 gal = _____ qt	3 qt 1 pt = _____ pt

Add, subtract, or multiply.

6.
```
   3 lb  6 oz
  +2 lb  8 oz
  _____
```
```
   8 ft  6 in.
  -2 ft  4 in.
  _____
```
```
   2 min  14 sec
           ×3
  _____
```

7.
```
   6 gal  2 qt
  -4 gal  3 qt
  _____
```
```
   3 yd  2 ft
          ×4
  _____
```
```
   6 hours 50 min
  +2 hours 48 min
  _____
```

Round as indicated.

	a nearest ten	*b* nearest hundred	*c* nearest thousand
8.	4773	_____	_____	_____
9.	63575	_____	_____	_____

Write an estimate for each exercise. Then find the answer.

10.
```
   7129
  +4516  _____
```
```
   9046
  -3978  _____
```
```
   296
  ×78  _____
```

Perfect score: 28   My score: _____

103

PRE-TEST—Geometry

a   b   c   d   e   f

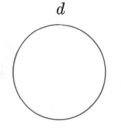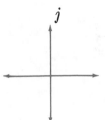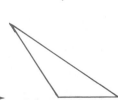

g   h   i   j   k   l

On the __ before each name below, write the letter(s) of the figure(s) it describes above.

a	b	c
1. _____ ray	_____ line segment	_____ isosceles triangle
2. _____ line	_____ obtuse angle	_____ obtuse triangle
3. _____ circle	_____ right triangle	_____ perpendicular lines
4. _____ acute angle	_____ parallel lines	_____ equilateral triangle
5. _____ right angle	_____ acute triangle	_____ scalene triangle

Use a protractor to find the measure of each angle below.

           a   b

6. _____°   _____°

Lesson 1   Lines, Line Segments, and Rays

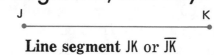

Line BC or $\overleftrightarrow{BC}$

Any two points on a line can be used to name that line. Do $\overleftrightarrow{BC}$ and $\overleftrightarrow{CB}$ name the same line? _____

_____

Line segment JK or $\overline{JK}$

$\overline{JK}$ consists of all points on the line between and including *endpoints* J and K. Do $\overline{JK}$ and $\overline{KJ}$ name the same line segment?

_____

Ray PQ or $\overrightarrow{PQ}$

$\overrightarrow{PQ}$ consists of endpoint P and all points on $\overrightarrow{PQ}$ which are on the same side of P as Q. Do $\overrightarrow{PQ}$ and $\overrightarrow{QP}$ name the same ray?

_____

Complete the following as shown.

1. E   D   <u>line ED or DE</u>   <u>$\overleftrightarrow{ED}$ or $\overleftrightarrow{DE}$</u>
           Endpoints: None

2. G   F   <u>ray FG</u>   <u>$\overrightarrow{FG}$</u>
           Endpoint: F

3. L   M   line segment LM or ML   <u>$\overline{LM}$ or $\overline{ML}$</u>
           Endpoints: L and M

4. R   S   _____   _____
           Endpoint(s): _____

5. X   Y   _____   _____
           Endpoint(s): _____

6. T   V   _____   _____
           Endpoint(s): _____

7. A   C   _____   _____
           Endpoint(s): _____

8. W   Z   _____   _____
           Endpoint(s): _____

9. H   N   _____   _____
           Endpoint(s): _____

Perfect score: 18   My score: _____

Lesson 2   Circles

By placing the compass point at point P, you can locate all the points in a plane (never-ending flat surface) which are the same distance from point P.

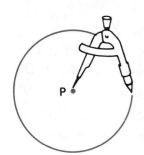

A circle is a set of points in a plane such that each point is the same distance from some given point called the *center*.

You can name a circle by naming its center. Circle P is shown at the left.

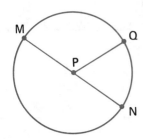

A radius of a circle is a line segment from the center of the circle to a point on the circle.

$\overline{PM}$ is a radius of circle P. Name two more radii of circle P. _____

A diameter of a circle is a line segment that has its endpoints on the circle and passes through the center of the circle.

Name a diameter of circle P. _____

Name the center, a radius, and a diameter of each circle.

1.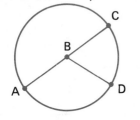

	center	*radius*	*diameter*
	_____	_____	_____

2.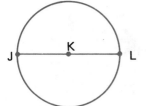

	_____	_____	_____

Write *True* or *False* after each statement.

3. All radii of the same circle have the same length. _____

4. All diameters of the same circle have the same length. _____

5. The length of a diameter of a circle is twice the length of a radius. _____

Perfect score: 9   My score: _____

Lesson 3   Angles

NAME _____

An **angle** is formed by two rays which have a common endpoint.

Study how angle ACB (denoted ∠ACB) is constructed below.

Step 1	*Step 2*	*Step 3*	
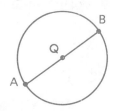	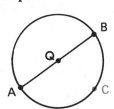	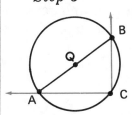	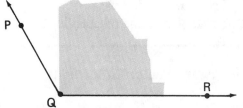
Draw circle Q and diameter AB.	Select point C anywhere on circle Q.	Draw $\overrightarrow{CA}$ and $\overrightarrow{CB}$.	Compare ∠ACB with a corner of a page of this book.

Angles such as ∠ACB are called **right angles**.

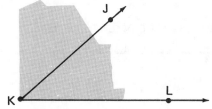

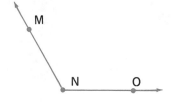

Does ∠JKL appear to be larger or smaller than a right angle? _____	Does ∠PQR appear to be larger or smaller than a right angle? _____
Angles like ∠JKL are called **acute angles**.	Angles like ∠PQR are called **obtuse angles**.

Compare each angle with a model of a right angle. Then describe each angle by writing either *acute*, *obtuse*, or *right* on each ___.

a	*b*	*c*
1. 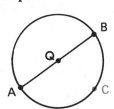	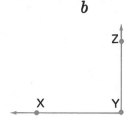	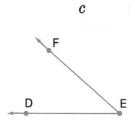
_____ angle	_____ angle	_____ angle
2. 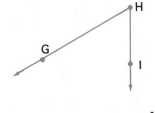	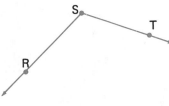	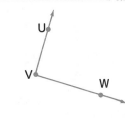
_____ angle	_____ angle	_____ angle

Perfect score: 6   My score: _____

107

Lesson 4   Angle Measurement

To use a protractor to measure an angle:

a. Place the center of the protractor at the vertex of the angle.

b. Align one side of the angle with the base of the protractor so that the other side of the angle intersects the curved edge of the protractor.

c. Use the scale starting at 0 and read the measure of the angle where the other side of the angle intersects the curved edge of the protractor.

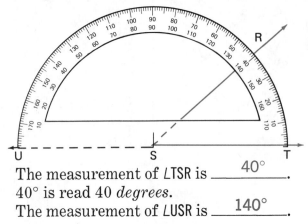

The measurement of ∠TSR is ___40°___.
40° is read 40 *degrees*.
The measurement of ∠USR is ___140°___.

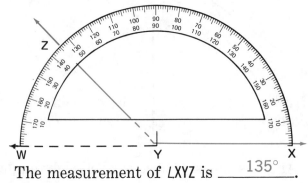

The measurement of ∠XYZ is ___135°___.

The measurement of ∠WYZ is _____.

Use a protractor to measure each angle below.

1.

a   *b*   *c*

_____°   _____°   _____°

2.

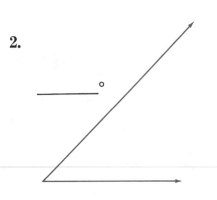

_____°   _____°   _____°

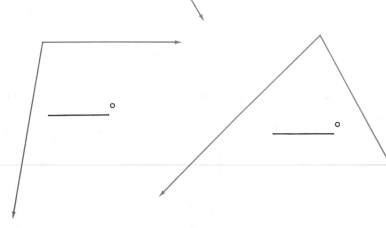

Perfect score: 6   My score: _____

NAME ______________________

Lesson 5 Angle Measurement

Use a protractor to measure each angle in the figure below.

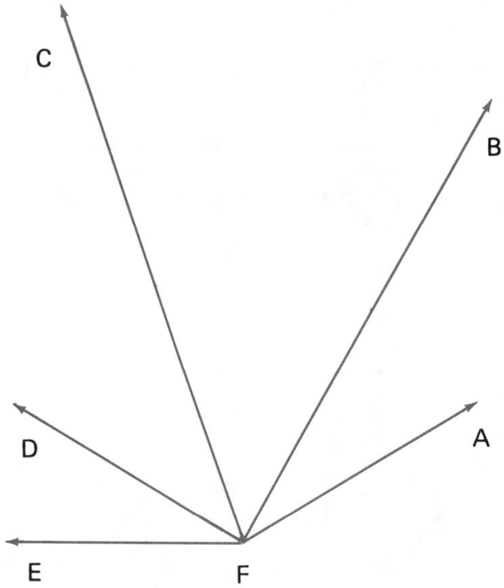

	a angle	*a* measurement	*b* angle	*b* measurement
1.	∠AFB	________	∠BFD	________
2.	∠BFC	________	∠CFE	________
3.	∠CFD	________	∠AFD	________
4.	∠DFE	________	∠BFE	________
5.	∠AFC	________	∠AFE	________

Use a protractor to draw angles having the measurements given below.

	a	*b*	*c*
6.	75°	90°	30°
7.	120°	60°	135°

Perfect score: 16 My score: ________

Lesson 6 Congruent Angles

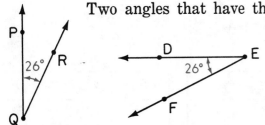

Two angles that have the same size are called **congruent angles.**

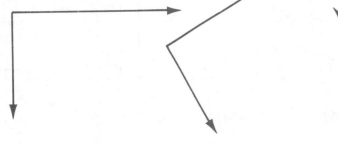

The measurement of ∠PQR is 26°.

The measurement of ∠DEF is 26°.

∠PQR ≅ ∠DEF (read ∠PQR *is congruent to* ∠DEF)

For each exercise, measure both angles. Write *congruent* if the angles are congruent. Write *not congruent* if the angles are not congruent.

a   b

1. _____   _____

2. _____   _____

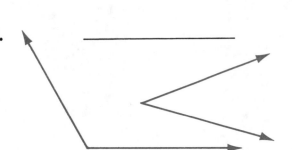
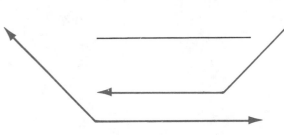

Find the measurement for each angle below. Then draw an angle congruent to each angle.

3.

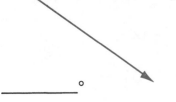

_____°   _____°

Perfect score: 6   My score: _____

Lesson 7   Parallel Lines and Perpendicular Lines

parallel lines   **perpendicular lines**

Parallel lines are always the same
distance apart. They will never in-
tersect, even if extended.

Perpendicular lines form
right angles.

Write *parallel* if the lines are parallel. Write *perpendicular* if the lines are perpendicular. Write *neither* if the lines are neither parallel nor perpendicular.

a   *b*   *c*

1. _____   _____   _____

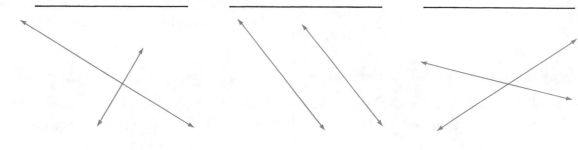

2. _____   _____   _____

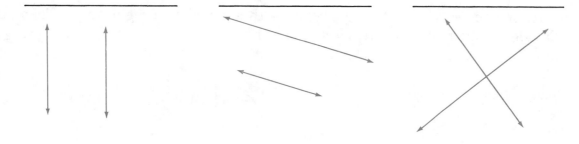

3. _____   _____   _____

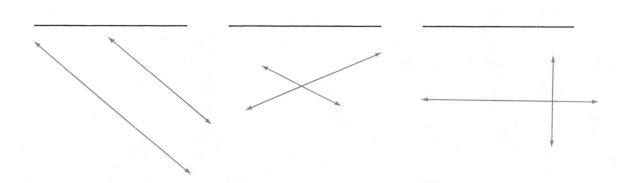

Perfect score: 9   My score: _____

Lesson 8 Triangles

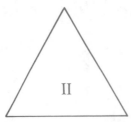

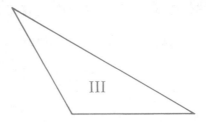

I

II

III

| Compare the angles of each triangle with a model of a right angle. | Use a compass to compare the lengths of the sides of each triangle. |

An **acute triangle** contains all acute angles.

Which triangle above is an acute triangle?

_____

A **right triangle** contains one right angle.

Which triangle above is a right triangle?

_____

An **obtuse triangle** contains one obtuse angle.

Which triangle above is an obtuse triangle?

_____

A **scalene triangle** has no sides the same length.

Which triangle above is a scalene triangle?

_____

An **isosceles triangle** has two or more sides the same length.

Which triangles above are isosceles triangles? _____

An **equilateral triangle** has all three sides the same length.

Which triangle above is an equilateral triangle? _____

Compare the angles of each triangle below with a model of a right angle. Then describe each triangle as being either *acute, obtuse,* or *right.*

a   *b*   *c*

1.

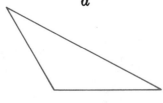

_____ triangle   _____ triangle   _____ triangle

2. Compare the lengths of the sides of each triangle. Then describe each triangle as being either *scalene, isosceles,* or *equilateral.*

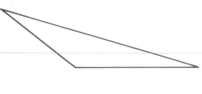

_____ triangle   _____ triangle   _____ triangle

Perfect score: 6   My score: _____

112

CHAPTER 8 TEST

Use the figures below to answer the questions which follow.

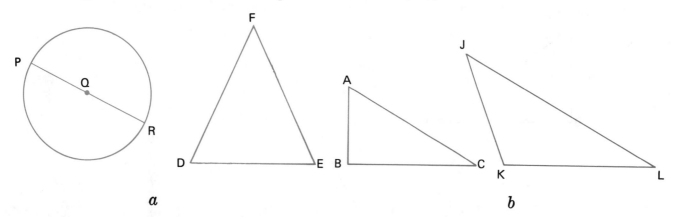

a   *b*

1. Name a radius of circle Q. _____   Name a diameter of circle Q. _____
2. Which figure is a right triangle?   Which figure is an isosceles triangle?

_____   _____

3. Which figures are scalene triangles?   Which figure is an obtuse triangle?

_____   _____

Use a protractor to measure each angle. Then describe each angle by writing *acute, obtuse,* or *right.*

a	*b*	*c*

4. _____°, _____   _____°, _____   _____°, _____

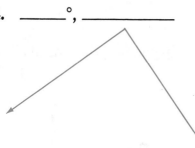  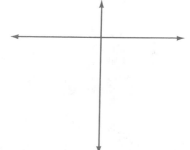

Tell whether each pair of lines is *parallel* or *perpendicular.*

5. _____   _____   _____

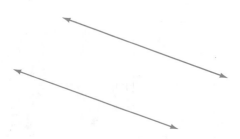

PRE-TEST—Similar Triangles and the Pythagorean Theorem

Find the length of the side shown in color in each pair of similar triangles below.

1.

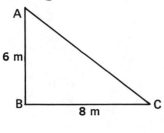

_____ m

2.

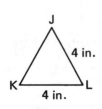

_____ in.

3.

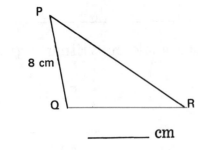

_____ cm

Use the Pythagorean Theorem and the table on page 121 to help you find the length of each side shown in color below.

4.

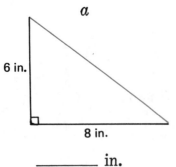

_____ in.   _____ m

5.

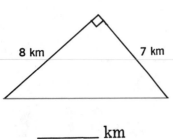

_____ km   _____ in.

Perfect score: 7   My score: _____

Lesson 1   Similar Triangles

When this photo was enlarged, the size of the angles did not change.

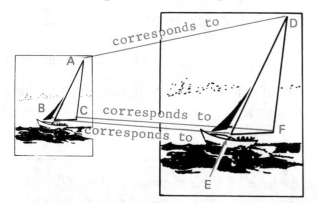

$\angle C \cong \angle F$

$\angle A \cong \angle D$

$\angle B \cong \angle E$

Did the lengths of the sides change? _____

Are the two triangles the same size? _____

$\triangle ABC$ *is similar to* $\triangle DEF$.

or

$\triangle ABC \sim \triangle DEF$

Two triangles are similar if their corresponding angles are congruent.

One triangle in each pair below is labeled with symbols like A′ (read A *prime*) and Q′ (read Q *prime*). This makes it easy to tell that A corresponds to A′ and Q corresponds to Q′, and so on. Complete the following.

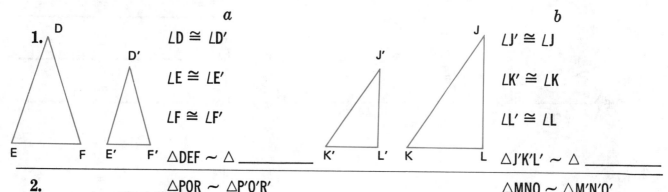

a

1.

$\angle D \cong \angle D'$

$\angle E \cong \angle E'$

$\angle F \cong \angle F'$

$\triangle DEF \sim \triangle$ _____

b

$\angle J' \cong \angle J$

$\angle K' \cong \angle K$

$\angle L' \cong \angle L$

$\triangle J'K'L' \sim \triangle$ _____

2. $\triangle PQR \sim \triangle P'Q'R'$

$\angle P \cong \angle$ _____

$\angle Q \cong \angle$ _____

$\angle R \cong \angle$ _____

$\triangle MNO \sim \triangle M'N'O'$

$\angle M \cong \angle$ _____

$\angle N \cong \angle$ _____

$\angle O \cong \angle$ _____

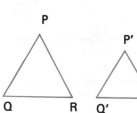

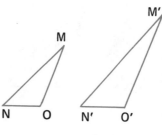

Write *True* or *False* after each statement below.

3. Two triangles that are similar must be the same size.   _____

4. Two triangles that are similar have the same shape.   _____

5. Two triangles that are similar have corresponding angles that are congruent.   _____

6. All right triangles are similar.   _____

Perfect score: 12   My score: _____

Lesson 2   Similar Triangles

NAME _____

△ABC~△A'B'C'

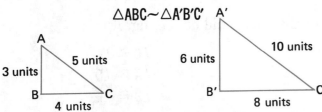

Side AB corresponds to side A'B'.

Side BC corresponds to side _____.

Side _____ corresponds to side C'A'.

If AB denotes the measure of side AB, A'B' the measure of side A'B', and so on, the ratios of the measures of corresponding sides can be expressed as follows.

$$\frac{AB}{A'B'} = \underline{\frac{3}{6}} = \underline{\frac{1}{2}} \qquad \frac{BC}{B'C'} = \underline{\frac{4}{8}} = \underline{\hspace{1cm}} \qquad \frac{CA}{C'A'} = \underline{\hspace{1cm}} = \underline{\hspace{1cm}}$$

> If two triangles are similar, the ratios of the measures of their corresponding sides are equal.

For each pair of similar triangles, complete the following to show that the ratios of the measures of corresponding sides are equal.

1.

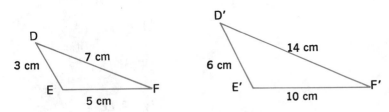

$$\frac{DE}{D'E'} = \underline{\hspace{1.5cm}} = \underline{\hspace{1.5cm}}$$

$$\frac{EF}{E'F'} = \underline{\hspace{1.5cm}} = \underline{\hspace{1.5cm}}$$

$$\frac{FD}{F'D'} = \underline{\hspace{1.5cm}} = \underline{\hspace{1.5cm}}$$

2.

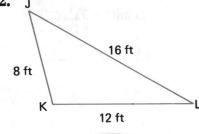

$$\frac{JK}{J'K'} = \underline{\hspace{1.5cm}} = \underline{\hspace{1.5cm}}$$

$$\frac{KL}{K'L'} = \underline{\hspace{1.5cm}} = \underline{\hspace{1.5cm}}$$

$$\frac{LJ}{L'J'} = \underline{\hspace{1.5cm}} = \underline{\hspace{1.5cm}}$$

3.

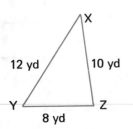

$$\frac{XY}{X'Y'} = \underline{\hspace{1.5cm}} = \underline{\hspace{1.5cm}}$$

$$\frac{YZ}{Y'Z'} = \underline{\hspace{1.5cm}} = \underline{\hspace{1.5cm}}$$

$$\frac{ZX}{Z'X'} = \underline{\hspace{1.5cm}} = \underline{\hspace{1.5cm}}$$

Perfect score: 18   My score: _____

Lesson 3 Similar Triangles

A 5-foot post casts a shadow 8 feet long while a nearby flagpole casts a shadow 48 feet long. What is the height of the flagpole?

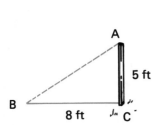

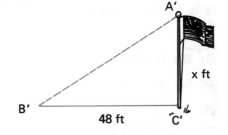

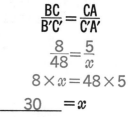

$$\frac{BC}{B'C'} = \frac{CA}{C'A'}$$

$$\frac{8}{48} = \frac{5}{x}$$

$$8 \times x = 48 \times 5$$

$$\underline{30} = x$$

The height of the flagpole is _____ feet.

Find the length of the side shown in color in each pair of similar triangles below.

a $\qquad\qquad\qquad\qquad\qquad\qquad\qquad$ b

1.

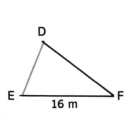

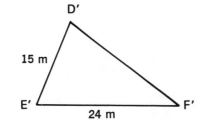

_____ m

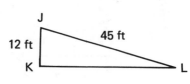

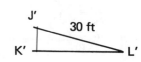

_____ ft

2.

_____ cm

_____ yd

3.

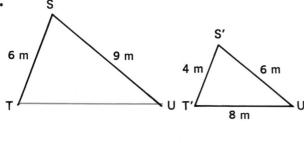

_____ m

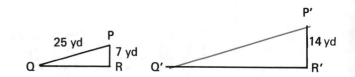

_____ in.

Perfect score: 6   My score: _____

117

Problem Solving

Solve each problem.

1. A tree 8 feet high casts a 4-foot shadow. At the same time, a nearby building casts a 16-foot shadow. What is the height of the building?

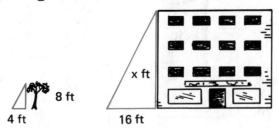

The height of the building is _____ feet.

2. If △CAB~△EDC, what is the length of the pond shown below?

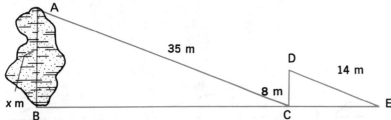

The length of the pond is _____ meters.

3. If △JKL~△PQL, what is the height of the building shown below?

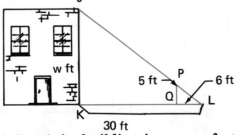

The height of the building is _____ feet.

4. A post 8 feet high casts a shadow 6 feet long. At the same time, a TV tower casts a shadow 30 feet long. How high is the TV tower?

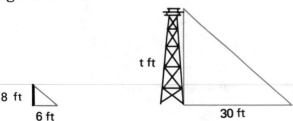

The TV tower is _____ feet high.

1.

2.

3.

4.

Perfect score: 4   My score: _____

Lesson 4   Squares and Square Roots

6^2 is read *6 squared*.

6^2 means 6×6.

$\sqrt{36}$ is read *the square root of 36*.

$\sqrt{36}$ is some number a so that $a \times a = 36$.

$6^2 = \underline{6} \times \underline{6} = \underline{36}$

$9^2 = \underline{9} \times \underline{9} = \underline{}$

$3^2 = \underline{} \times \underline{} = \underline{}$

$\sqrt{36} = \sqrt{\underline{6} \times \underline{6}} = \underline{6}$

$\sqrt{81} = \sqrt{\underline{9} \times \underline{9}} = \underline{}$

$\sqrt{9} = \sqrt{\underline{} \times \underline{}} = \underline{}$

Complete the following.

	a		*b*

1. $5^2 = \underline{} \times \underline{} = \underline{}$ $\sqrt{25} = \sqrt{\underline{} \times \underline{}} = \underline{}$

2. $8^2 = \underline{} \times \underline{} = \underline{}$ $\sqrt{64} = \sqrt{\underline{} \times \underline{}} = \underline{}$

3. $2^2 = \underline{} \times \underline{} = \underline{}$ $\sqrt{4} = \sqrt{\underline{} \times \underline{}} = \underline{}$

4. $10^2 = \underline{} \times \underline{} = \underline{}$ $\sqrt{100} = \sqrt{\underline{} \times \underline{}} = \underline{}$

5. $4^2 = \underline{} \times \underline{} = \underline{}$ $\sqrt{16} = \sqrt{\underline{} \times \underline{}} = \underline{}$

6. $12^2 = \underline{} \times \underline{} = \underline{}$ $\sqrt{144} = \sqrt{\underline{} \times \underline{}} = \underline{}$

7. $20^2 = \underline{} \times \underline{} = \underline{}$ $\sqrt{400} = \sqrt{\underline{} \times \underline{}} = \underline{}$

8. $11^2 = \underline{} \times \underline{} = \underline{}$ $\sqrt{121} = \sqrt{\underline{} \times \underline{}} = \underline{}$

9. $19^2 = \underline{} \times \underline{} = \underline{}$ $\sqrt{361} = \sqrt{\underline{} \times \underline{}} = \underline{}$

10. $25^2 = \underline{} \times \underline{} = \underline{}$ $\sqrt{625} = \sqrt{\underline{} \times \underline{}} = \underline{}$

11. $31^2 = \underline{} \times \underline{} = \underline{}$ $\sqrt{961} = \sqrt{\underline{} \times \underline{}} = \underline{}$

12. $43^2 = \underline{} \times \underline{} = \underline{}$ $\sqrt{1849} = \sqrt{\underline{} \times \underline{}} = \underline{}$

13 $50^2 = \underline{} \times \underline{} = \underline{}$ $\sqrt{2500} = \sqrt{\underline{} \times \underline{}} = \underline{}$

Perfect score: 26   My score: _____

Lesson 5 Squares and Square Roots

Study how the table is used to find the square and the square root of a number n. ($\approx$ is read *is approximately equal to*.)

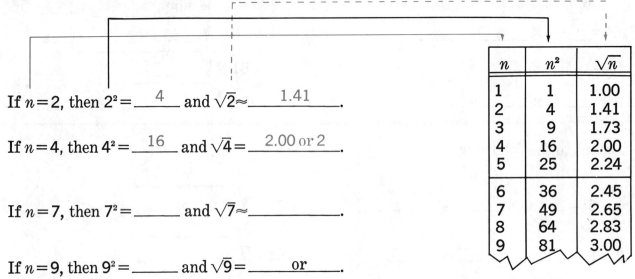

n	n^2	$\sqrt{n}$
1	1	1.00
2	4	1.41
3	9	1.73
4	16	2.00
5	25	2.24
6	36	2.45
7	49	2.65
8	64	2.83
9	81	3.00

If $n=2$, then $2^2 = $ __4__ and $\sqrt{2} \approx$ __1.41__ .

If $n=4$, then $4^2 = $ __16__ and $\sqrt{4} = $ __2.00 or 2__ .

If $n=7$, then $7^2 = $ _____ and $\sqrt{7} \approx$ _____ .

If $n=9$, then $9^2 = $ _____ and $\sqrt{9} = $ _____ or ____ .

Use the table on page 121 to help you complete the following.

1. If $n=18$, then $18^2 = $ _____ and $\sqrt{18} \approx$ _____ .

2. If $n=25$, then $25^2 = $ _____ and $\sqrt{25} = $ _____ .

3. If $n=45$, then $45^2 = $ _____ and $\sqrt{45} \approx$ _____ .

4. If $n=64$, then $64^2 = $ _____ and $\sqrt{64} = $ _____ .

5. If $n=83$, then $83^2 = $ _____ and $\sqrt{83} \approx$ _____ .

6. If $n=75$, then $75^2 = $ _____ and $\sqrt{75} \approx$ _____ .

7. If $n=90$, then $90^2 = $ _____ and $\sqrt{90} \approx$ _____ .

8. If $n=104$, then $104^2 = $ _____ and $\sqrt{104} \approx$ _____ .

9. If $n=135$, then $135^2 = $ _____ and $\sqrt{135} \approx$ _____ .

10. If $n=147$, then $147^2 = $ _____ and $\sqrt{147} \approx$ _____ .

11. If $n=150$, then $150^2 = $ _____ and $\sqrt{150} \approx$ _____ .

Perfect score: 22   My score: _____

Lesson 6   Squares and Square Roots

Table of Squares and Square Roots								
n	n^2	$\sqrt{n}$	n	n^2	$\sqrt{n}$	n	n^2	$\sqrt{n}$
1	1	1.00	51	2,601	7.14	101	10,201	10.05
2	4	1.41	52	2,704	7.21	102	10,404	10.10
3	9	1.73	53	2,809	7.28	103	10,609	10.15
4	16	2.00	54	2,916	7.35	104	10,816	10.20
5	25	2.24	55	3,025	7.42	105	11,025	10.25
6	36	2.45	56	3,136	7.48	106	11,236	10.30
7	49	2.65	57	3,249	7.55	107	11,449	10.34
8	64	2.83	58	3,364	7.62	108	11,664	10.39
9	81	3.00	59	3,481	7.68	109	11,881	10.44
10	100	3.16	60	3,600	7.75	110	12,100	10.49
11	121	3.32	61	3,721	7.81	111	12,321	10.54
12	144	3.46	62	3,844	7.87	112	12,544	10.58
13	169	3.61	63	3,969	7.94	113	12,769	10.63
14	196	3.74	64	4,096	8.00	114	12,996	10.68
15	225	3.87	65	4,225	8.06	115	13,225	10.72
16	256	4.00	66	4,356	8.12	116	13,456	10.77
17	289	4.12	67	4,489	8.19	117	13,689	10.82
18	324	4.24	68	4,624	8.25	118	13,924	10.86
19	361	4.36	69	4,761	8.31	119	14,161	10.91
20	400	4.47	70	4,900	8.37	120	14,400	10.95
21	441	4.58	71	5,041	8.43	121	14,641	11.00
22	484	4.69	72	5,184	8.49	122	14,884	11.05
23	529	4.80	73	5,329	8.54	123	15,129	11.09
24	576	4.90	74	5,476	8.60	124	15,376	11.14
25	625	5.00	75	5,625	8.66	125	15,625	11.18
26	676	5.10	76	5,776	8.72	126	15,876	11.22
27	729	5.20	77	5,929	8.77	127	16,129	11.27
28	784	5.29	78	6,084	8.83	128	16,384	11.31
29	841	5.39	79	6,241	8.89	129	16,641	11.36
30	900	5.48	80	6,400	8.94	130	16,900	11.40
31	961	5.57	81	6,561	9.00	131	17,161	11.45
32	1,024	5.66	82	6,724	9.06	132	17,424	11.49
33	1,089	5.74	83	6,889	9.11	133	17,689	11.53
34	1,156	5.83	84	7,056	9.17	134	17,956	11.58
35	1,225	5.92	85	7,225	9.22	135	18,225	11.62
36	1,296	6.00	86	7,396	9.27	136	18,496	11.66
37	1,369	6.08	87	7,569	9.33	137	18,769	11.70
38	1,444	6.16	88	7,744	9.38	138	19,044	11.75
39	1,521	6.24	89	7,921	9.43	139	19,321	11.79
40	1,600	6.32	90	8,100	9.49	140	19,600	11.83
41	1,681	6.40	91	8,281	9.54	141	19,881	11.87
42	1,764	6.48	92	8,464	9.59	142	20,164	11.92
43	1,849	6.56	93	8,649	9.64	143	20,449	11.96
44	1,936	6.63	94	8,836	9.70	144	20,736	12.00
45	2,025	6.71	95	9,025	9.75	145	21,025	12.04
46	2,116	6.78	96	9,216	9.80	146	21,316	12.08
47	2,209	6.86	97	9,409	9.85	147	21,609	12.12
48	2,304	6.93	98	9,604	9.90	148	21,904	12.17
49	2,401	7.00	99	9,801	9.95	149	22,201	12.21
50	2,500	7.07	100	10,000	10.00	150	22,500	12.25

Square Roots

Study how the table on page **121** can be used to find the square root of a number greater than 150.

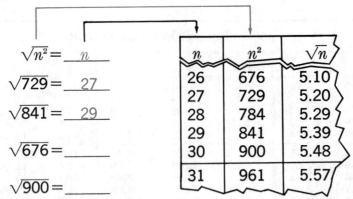

$$\sqrt{n^2} = \underline{\quad n \quad}$$

$$\sqrt{729} = \underline{\quad 27 \quad}$$

$$\sqrt{841} = \underline{\quad 29 \quad}$$

$$\sqrt{676} = \underline{\qquad}$$

$$\sqrt{900} = \underline{\qquad}$$

n	n^2	$\sqrt{n}$
26	676	5.10
27	729	5.20
28	784	5.29
29	841	5.39
30	900	5.48
31	961	5.57

Use the table on page **121** to help you complete each of the following.

	a	b	c
1.	$\sqrt{169} = \underline{\qquad}$	$\sqrt{529} = \underline{\qquad}$	$\sqrt{784} = \underline{\qquad}$
2.	$\sqrt{256} = \underline{\qquad}$	$\sqrt{361} = \underline{\qquad}$	$\sqrt{961} = \underline{\qquad}$
3.	$\sqrt{1225} = \underline{\qquad}$	$\sqrt{2209} = \underline{\qquad}$	$\sqrt{3969} = \underline{\qquad}$
4.	$\sqrt{1681} = \underline{\qquad}$	$\sqrt{3136} = \underline{\qquad}$	$\sqrt{4761} = \underline{\qquad}$
5.	$\sqrt{5329} = \underline{\qquad}$	$\sqrt{6084} = \underline{\qquad}$	$\sqrt{6889} = \underline{\qquad}$
6.	$\sqrt{7921} = \underline{\qquad}$	$\sqrt{8649} = \underline{\qquad}$	$\sqrt{9604} = \underline{\qquad}$
7.	$\sqrt{10201} = \underline{\qquad}$	$\sqrt{11025} = \underline{\qquad}$	$\sqrt{11449} = \underline{\qquad}$
8.	$\sqrt{12544} = \underline{\qquad}$	$\sqrt{17424} = \underline{\qquad}$	$\sqrt{22201} = \underline{\qquad}$
9.	$\sqrt{4900} = \underline{\qquad}$	$\sqrt{1849} = \underline{\qquad}$	$\sqrt{16900} = \underline{\qquad}$
10.	$\sqrt{10000} = \underline{\qquad}$	$\sqrt{12100} = \underline{\qquad}$	$\sqrt{2500} = \underline{\qquad}$
11.	$\sqrt{8464} = \underline{\qquad}$	$\sqrt{19321} = \underline{\qquad}$	$\sqrt{13924} = \underline{\qquad}$
12.	$\sqrt{8281} = \underline{\qquad}$	$\sqrt{18225} = \underline{\qquad}$	$\sqrt{21316} = \underline{\qquad}$

Perfect score: 36   My score: _____

Lesson 7   The Pythagorean Theorem

In a right triangle the side opposite the right angle is called the **hypotenuse.**

⌐ means a right angle.

Which side is the hypotenuse in $\triangle PQR$?_____

Pythagorean Theorem: The square of the measure of the hypotenuse of a right triangle is equal to the sum of the squares of the measures of the other two sides.

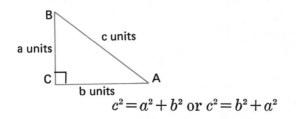

$$c^2 = a^2 + b^2 \text{ or } c^2 = b^2 + a^2$$

Find c if $a = 3$ and $b = 4$.

$$c^2 = a^2 + b^2$$
$$c^2 = 3^2 + 4^2$$
$$c^2 = 9 + 16 \text{ or } 25$$
$$\sqrt{c^2} = \sqrt{25}$$
$$c = \underline{\quad}$$

Use $\triangle ABC$ above and the table on page 121 to help you complete the following.

1. If $a = 6$ and $b = 8$, then $c = $ _____.

2. If $a = 7$ and $b = 24$, then $c = $ _____.

3. If $a = 5$ and $b = 7$, then $c \approx $ _____.

4. If $a = 7$ and $b = 9$, then $c \approx $ _____.

5. If $a = 5$ and $b = 12$, then $c = $ _____.

6. If $a = 8$ and $b = 8$, then $c \approx $ _____.

7. If $a = 8$ and $b = 15$, then $c = $ _____.

8. If $a = 3$ and $b = 8$, then $c \approx $ _____.

9. If $a = 20$ and $b = 21$, then $c = $ _____.

10. If $a = 12$ and $b = 2$, then $c \approx $ _____.

11. If $a = 45$ and $b = 28$, then $c = $ _____.

Perfect score: 11   My score: _____

Problem Solving

Use the Pythagorean Theorem and the table on page **121** to help you solve each of the following.

1. The foot of a ladder is placed 5 feet from a building. The top of the ladder rests 12 feet up on the building. How long is the ladder?

The ladder is _____ feet long.

1.

2. The roof of a house is to be built as shown. How long should each rafter be?

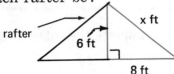

Each rafter should be _____ feet long.

2.

3. A ship left port and sailed 5 miles east and then 7 miles north. How far was the ship from the port then?

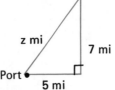

The ship was about _____ miles from the port.

3.

4. What is the length of the lake shown below?

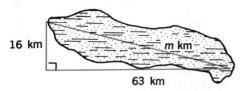

The length is _____ kilometers.

4.

5. An inclined ramp rises 5 meters over a horizontal distance of 11 meters. What is the length of the ramp?

The length is _____ meters.

5.

Perfect score: 5   My score: _____

Lesson 8   Using the Pythagorean Theorem

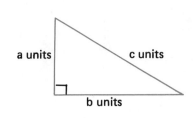

a units

c units

b units

Find a if $c = 17$ and $b = 15$.

$$c^2 = a^2 + b^2$$
$$17^2 = a^2 + 15^2$$
$$289 = a^2 + 225$$
$$289 - 225 = a^2 + 225 - 225$$
$$64 = a^2$$
$$\sqrt{64} = \sqrt{a^2}$$
$$\underline{\quad 8 \quad} = a$$

Find b if $c = 13$ and $a = 12$.

$$c^2 = b^2 + a^2$$
$$13^2 = b^2 + 12^2$$
$$169 = b^2 + 144$$
$$169 - 144 = b^2 + 144 - 144$$
$$25 = b^2$$
$$\sqrt{25} = \sqrt{b^2}$$
$$\underline{\quad\quad} = b$$

Use the triangle above and the table on page **121** to help you complete the following.

1. If $c = 25$ and $a = 24$, then $b = $_____.

2. If $c = 41$ and $b = 40$, then $a = $_____.

3. If $c = 61$ and $a = 60$, then $b = $_____.

4. If $c = 25$ and $b = 22$, then $a \approx$_____.

5. If $c = 26$ and $a = 24$, then $b = $_____.

6. If $c = 20$ and $b = 17$, then $a \approx$_____.

7. If $c = 89$ and $a = 80$, then $b = $_____.

8. If $c = 65$ and $b = 63$, then $a = $_____.

9. If $c = 72$ and $a = 71$, then $b \approx$_____.

10. If $c = 73$ and $b = 48$, then $a = $_____.

11. If $c = 38$ and $a = 36$, then $b \approx$_____.

12. If $c = 85$ and $b = 36$, then $a = $_____.

Perfect score: 12   My score: _____

Problem Solving

Use the Pythagorean Theorem and the table on page **121** to help you solve each of the following.

1. Suppose the foot of a 12-foot ladder was placed 4 feet from the building. How high up on the building would the top of the ladder reach?

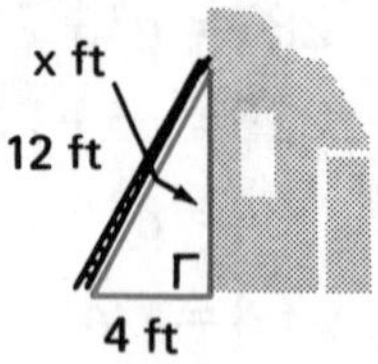

The ladder will reach about _____ feet.

2. A ship is 24 kilometers east of port. How far north must the ship sail to reach a point that is 25 kilometers from the port?

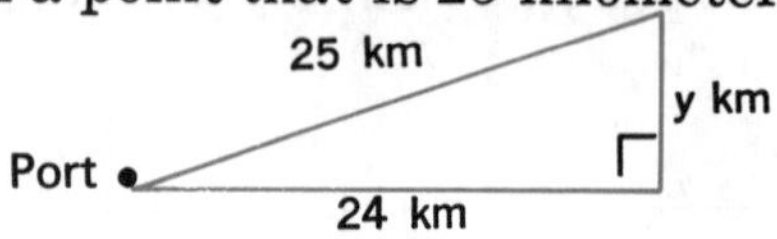

The ship must sail _____ kilometers north.

3. A telephone pole is braced by a guy wire as shown below. How high up on the pole is the wire fastened?

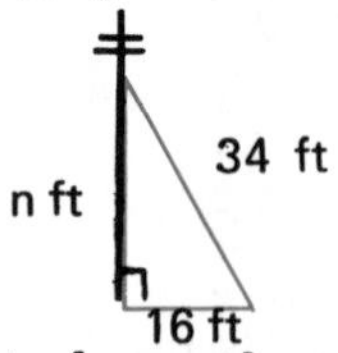

The wire is fastened _____ feet above the ground.

4. How far is it across the pond shown below?

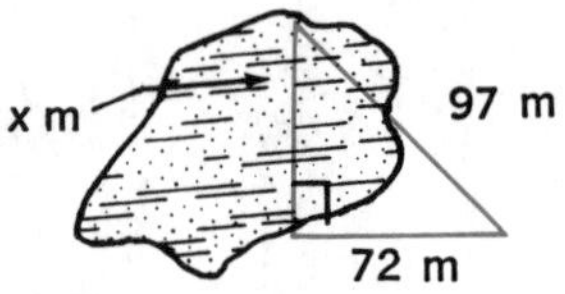

It is _____ meters across the pond.

5. Alstown, Donville, and Maxburg are located as shown below. How many miles is it from Alstown to Maxburg?

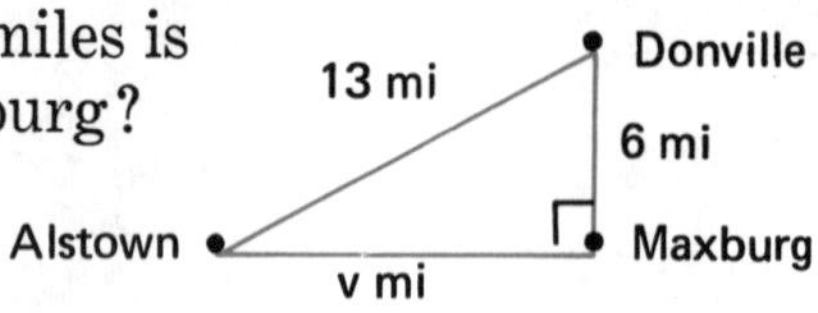

It is about _____ miles from Alstown to Maxburg.

6. A sail is shaped as shown. How high is the sail?

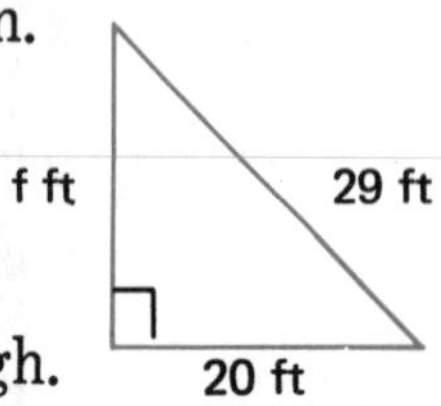

The sail is _____ feet high.

1.

2.

3.

4.

5.

6.

Perfect score: 6   My score: _____

126

Lesson 9   Similar Right Triangles

Study how the Pythagorean Theorem and the ratios of similar triangles are used to find the measure of $\overline{AB}$, the measure of $\overline{A'C'}$, and the measure of $\overline{B'C'}$.

right $\triangle ABC \sim$ right $\triangle A'B'C'$

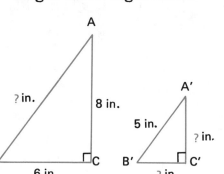

Step 1

Use $c^2 = a^2 + b^2$ to find the measure of $\overline{AB}$.

$$c^2 = a^2 + b^2$$
$$c^2 = 6^2 + 8^2$$
$$c^2 = 100$$
$$c = 10$$
$$AB = \underline{10}$$

Step 2

Use the measure of $\overline{AB}$ from *Step 1* and find the measure of $\overline{A'C'}$ and $\overline{B'C'}$.

$$\frac{AB}{A'B'} = \frac{AC}{A'C'} \qquad \frac{AB}{A'B'} = \frac{BC}{B'C'}$$
$$\frac{10}{5} = \frac{8}{A'C'} \qquad \frac{10}{5} = \frac{6}{B'C'}$$
$$A'C' = \underline{4} \qquad B'C' = \underline{3}$$

$\overline{AB}$ is _____ inches long. $\overline{A'C'}$ is _____ inches long. $\overline{B'C'}$ is _____ inches long.

Find the length of each side shown in color in each pair of similar right triangles below. You may use the table on page 121 if necessary.

1.

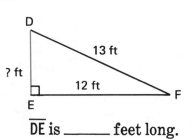

$\overline{DE}$ is _____ feet long. $\overline{E'F'}$ is _____ feet long. $\overline{D'F'}$ is _____ feet long.

2.

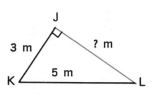

$\overline{JL}$ is _____ meters long. $\overline{J'K'}$ is _____ meters long. $\overline{K'L'}$ is _____ meters long.

3.

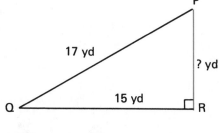

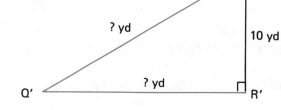

$\overline{PR}$ is _____ yards long. $\overline{Q'R'}$ is _____ yards long. $\overline{Q'P'}$ is _____ yards long.

Perfect score: 9   My score: _____

Problem Solving

Solve each problem. If necessary use the table on page **121**.

1. If △ABD∼△ECD, how far is it from the pier to the island? From the boathouse to the campsite? From the lodge to the campsite?

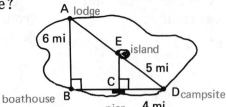

It is _____ miles from the pier to the island.

It is _____ miles from the boathouse to the campsite.

It is _____ miles from the lodge to the campsite.

2. A telephone pole is steadied by guy wires as shown. If △JKL∼△MKN, what are the lengths of the guy wires? How high is the upper guy wire fastened above the ground?

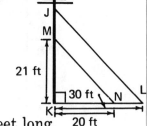

The shorter wire is _____ feet long.

The longer wire is _____ feet long.

The upper guy wire is fastened _____ feet above the ground.

3. Two inclined ramps are shaped as shown below and △PQR∼△XYZ. What is the length of QR? What is the height of each ramp?

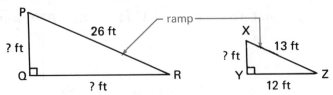

The length of the taller ramp is _____ feet.

The height of the shorter ramp is _____ feet.

The height of the taller ramp is _____ feet.

| 1. |
| 2. |
| 3. |

Perfect score: 9   My score: _____

128

CHAPTER 9 TEST

Use the similar triangles below to help you complete the following.

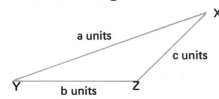
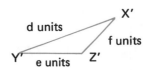

$$\frac{a}{d} = \frac{b}{e} = \frac{c}{f}$$

1. If $a = 6$, $d = 9$,
and $b = 8$, then $e =$ _____.

2. If $b = 24$, $e = 12$,
and $c = 16$, then $f =$ _____.

3. If $c = 8$, $f = 4$,
and $e = 6$, then $b =$ _____.

4. If $e = 9$, $b = 3$,
and $d = 6$, then $a =$ _____.

Use the triangle below and the table on page **121** to help you complete the following.

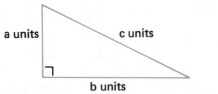

$$c^2 = a^2 + b^2 \text{ or } c^2 = b^2 + a^2$$

5. If $a = 8$ and $b = 6$, then $c =$ _____.

6. If $a = 16$ and $c = 65$, then $b =$ _____.

7. If $b = 20$ and $c = 23$, then $a \approx$ _____.

8. If $a = 9$ and $b = 8$, then $c \approx$ _____.

Solve each of the following.

9. A post and a flagpole cast shadows as shown below. What is the height of the flagpole?

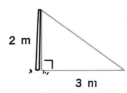
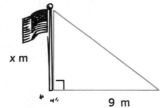

2 m   x m

3 m   9 m

The height of the flagpole is _____ meters.

10. A windowpane is 7 inches by 7 inches. What is the distance between opposite corners of the windowpane?

The distance is about _____ inches.

Perfect score: 10   My score: _____

PRE-TEST Perimeter, Area, and Volume

Find the perimeter or circumference and the area of each figure below. Use $3\frac{1}{7}$ for π.

| a | b | c |

1.

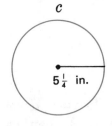

5 m

7 m

7 cm 8 cm 5 cm 9 cm

$5\frac{1}{4}$ in.

perimeter: _____ meters   _____ centimeters   _____ inches

area: _____ square meters   _____ square centimeters   _____ square inches

2.

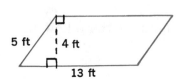

5 ft 4 ft 13 ft

5 m 6 m 8 m

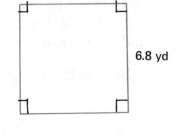

6.8 yd

perimeter: _____ feet   _____ meters   _____ yards

area: _____ square feet   _____ square meters   _____ square yards

Find the volume of each figure below. Use 3.14 for π.

| a | b |

3.

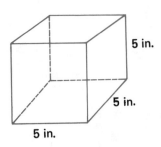

5 in. 5 in. 5 in.

3.7 cm 4.6 cm 8.4 cm

_____ cubic inches   _____ cubic centimeters

4.

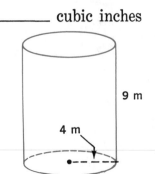

9 m 4 m

6 in. 4 in. 5 in.

_____ cubic meters   _____ cubic inches

Perfect score: 16   My score: _____

Lesson 1   Perimeter

The perimeter measure (p) of a figure is equal to the sum of the measures of its sides.

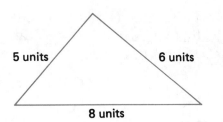

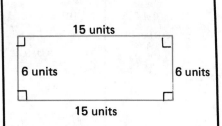

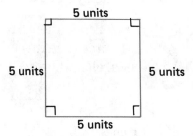

Find p if $a=5$, $b=6$, and $c=8$.

$$p=a+b+c$$
$$=5+8+6$$
$$=\underline{\ 19\ }$$

Find p if $l=15$ and $w=6$.

$$p=l+w+l+w$$
$$=2(l+w)$$
$$=2(15+6)$$
$$=2\times21 \text{ or } \underline{\quad}$$

Find p if $s=5$.

$$p=s+s+s+s$$
$$=4s$$
$$=4\times5$$
$$=\underline{\quad}$$

The perimeter is _19_ units. | The perimeter is ____ units. | The perimeter is ____ units.

Find the perimeter of each figure below.

| | a | b | c |

1.

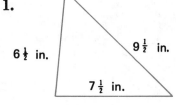

6½ in.   9½ in.   7½ in.

_____ in.

13 ft   7 ft

_____ ft

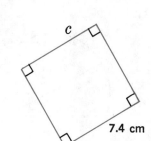

7.4 cm

_____ cm

2.

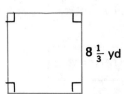

8⅓ yd

_____ yd

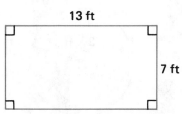

6.7 m   4.2 m

_____ m

7 in.   8 in.   9 in.   8 in.   10 in.

_____ in.

3.

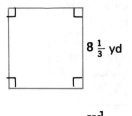

6½ ft   9¼ ft

_____ ft

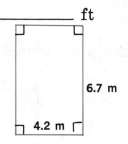

7.6 cm   13.2 cm

_____ cm

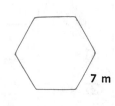

7 m

_____ m

Perfect score: 9   My score: _____

Lesson 2   Circumference

The ratio of the measure of the circumference to the measure of a diameter is the same for all circles. The symbol π stands for this ratio. π is approximately equal to 3.14 or $3\frac{1}{7}$.

| The circumference measure (C) of a circle is equal to π times the measure of a diameter (d) of the circle. $C = \pi d$ | The measure of a diameter (d) is twice the measure of a radius (r). Hence, $C = \pi d$ can be changed to $C = \pi(2r)$ or $C = 2\pi r$. |

Find C if $d = 7$.

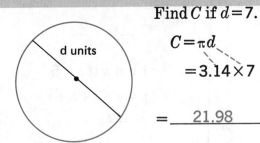

$$C = \pi d$$
$$= 3.14 \times 7$$
$$= \underline{21.98}$$

The circumference is _____ units.

Find C if $r = 6$.

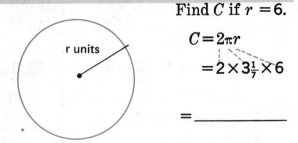

$$C = 2\pi r$$
$$= 2 \times 3\frac{1}{7} \times 6$$
$$= \underline{\hspace{2cm}}$$

The circumference is _____ units.

Find the circumference of each circle below. Use $3\frac{1}{7}$ for π.

| a | b | c |

1.

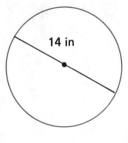

14 in

_____ in.

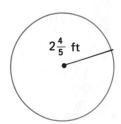

$2\frac{4}{5}$ ft

_____ ft

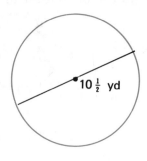

$10\frac{1}{2}$ yd

_____ yd

Find the circumference of each circle described below. Use 3.14 for π.

a		b	
diameter	circumference	radius	circumference
2. 6 m	_____ m	21 mm	_____ mm
3. 15 cm	_____ cm	6.7 cm	_____ cm
4. 6.8 km	_____ km	48 cm	_____ cm
5. 81 mm	_____ mm	37 mm	_____ mm
6. 27 mm	_____ mm	9.6 m	_____ m
7. 4.2 m	_____ m	4 km	_____ km

Perfect score: 15   My score: _____

132

Lesson 3   Area of a Rectangle

The area measure (A) of a rectangle is equal to the product of the measure of its length (l) and the measure of its width (w). $A = l \times w$ or $A = lw$

Find A if $l=9$ and $w=5$.

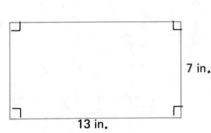

$$A = lw$$
$$= 9 \times 5$$
$$= \underline{45}$$

The area is _____ square units.

Find A if $s=3$.

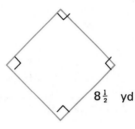

$$A = s \times s \text{ or } s^2$$
$$= 3 \times 3$$
$$= \underline{}$$

The area is _____ square units.

Find the area of each rectangle below.

          a b c

1.

7 in.

13 in.

$8\frac{1}{2}$ yd

9.5 m

6.5 m

_____ square inches   _____ square yards   _____ square meters

Find the area of each rectangle described below.

	length	width	area
2.	33 in.	27 in.	_____ square inches
3.	$5\frac{1}{4}$ ft	$3\frac{1}{2}$ ft	_____ square feet
4.	$3\frac{3}{4}$ yd	2 yd	_____ square yards
5.	6.7 m	6.7 m	_____ square meters
6.	9.2 cm	7.7 cm	_____ square centimeters
7.	18 m	4.6 m	_____ square meters
8.	3.6 km	3.6 km	_____ square kilometers
9.	9.5 cm	6.6 cm	_____ square centimeters

Perfect score: 11   My score: _____

Lesson 4   Area of a Triangle

The area measure (A) of a triangle is equal to $\frac{1}{2}$ the product of the measure of its base (b) and the measure of its height (h). $A = \frac{1}{2}bh$ or $A = .5bh$

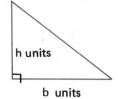

Find A if $b=8$ and $h=6$.

$A = \frac{1}{2}bh$

$= \frac{1}{2} \times 8 \times 6$

$= \underline{\ 24\ }$

The area is _____ square units.

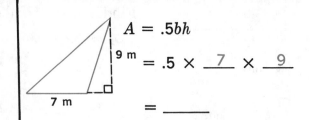

$A = .5bh$

$= .5 \times \underline{\ 7\ } \times \underline{\ 9\ }$

$= \underline{\qquad}$

The area is _____ square meters.

Find the area of each triangle below.

| a | b | c |

1.

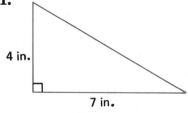

4 in.

7 in.

$3\frac{1}{2}$ ft

$7\frac{1}{2}$ ft

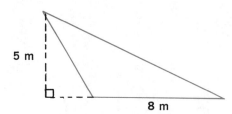

5 m

8 m

_____ square inches   _____ square feet   _____ square meters

Find the area of each triangle described below.

	base	height	area
2.	15 ft	9 ft	_____ square feet
3.	$3\frac{1}{2}$ in.	$6\frac{1}{2}$ in.	_____ square inches
4.	7.4 cm	6.5 cm	_____ square centimeters
5.	$11\frac{1}{2}$ yd	7 yd	_____ square yards
6.	154 mm	37 mm	_____ square millimeters
7.	85 cm	35 cm	_____ square centimeters
8.	18.8 m	7.5 m	_____ square meters
9.	9.5 km	6.6 km	_____ square kilometers

Perfect score: 11   My score: _____

Lesson 5   Area of a Circle

The area measure (A) of a circle is equal to the product of π and the square of the measure of a radius (r^2) of the circle. $A = \pi r^2$

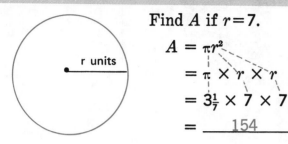

Find A if $r=7$.

$A = \pi r^2$

$\quad = \pi \times r \times r$

$\quad = 3\frac{1}{7} \times 7 \times 7$

$\quad = \underline{\quad 154 \quad}$

The area is _____ square units.

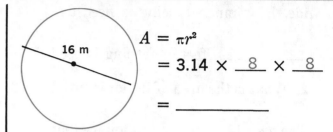

16 m

$A = \pi r^2$

$\quad = 3.14 \times \underline{\ 8\ } \times \underline{\ 8\ }$

$\quad = \underline{\qquad\qquad}$

The area is _____ square meters.

Find the area of each circle below. Use 3.14 for π.

| a | b | c |

1.

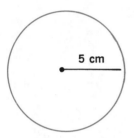

5 cm

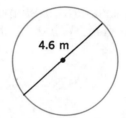

4.6 m

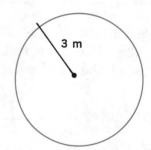

3 m

_____ square centimeters   _____ square meters   _____ square meters

Find the area of each circle described below. Use $3\frac{1}{7}$ for π.

a

radius	area
2. 9 ft	_____ square feet
3. 14 in.	_____ square inches
4. $3\frac{1}{2}$ yd	_____ square yards
5. 56 ft	_____ square feet
6. $5\frac{1}{4}$ in.	_____ square inches
7. 45 yd	_____ square yards

b

diameter	area
28 in.	_____ square inches
42 ft	_____ square feet
72 yd	_____ square yards
126 in.	_____ square inches
84 ft	_____ square feet
$1\frac{3}{4}$ in.	_____ square inches

Perfect score: 15   My score: _____

135

Problem Solving

Solve each problem. Use $3\frac{1}{7}$ for π.

1. The Redfords would like to build a fence around a rectangular lot. The lot is 140 feet long and 50 feet wide. How much fencing is needed?

_____ feet of fencing is needed.

2. What is the area of the lot in problem 1?

The area is _____ square feet.

3. Mrs. McDaniel wants to put carpeting in a room that is 15 feet long and 12 feet wide. How many square feet of carpeting does she need?

She needs _____ square feet of carpeting.

4. The lengths of the sides of a triangular-shaped garden are 17 feet, 26 feet, and 35 feet. What is the perimeter of the garden?

The perimeter is _____ feet.

5. The length of a diameter of a circular pond is 28 feet. What is the circumference of the pond?
(Use $3\frac{1}{7}$ for π.)

The circumference is _____ feet.

6. What is the area of the pond in problem 5?

The area is _____ square feet.

7. Mr. Witt is refinishing a circular table with a radius of 60 centimeters. Find the area of the table-top. (Use 3.14 for π.)

The area is _____ square centimeters.

8. Find the circumference of the tabletop in problem 7.

The circumference is _____ centimeters.

1.	2.
3.	4.
5.	6.
7.	8.

Perfect score: 8   My score: _____

136

Lesson 6 Area of a Parallelogram

The area measure (A) of a parallelogram is equal to the product of the measure of its base (b) and the measure of its height (h). $A = bh$

Find A if $b = 14$ and $h = 9$.

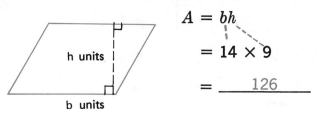

$A = bh$

$= 14 \times 9$

$= \underline{\quad 126 \quad}$

The area is _____ square units.

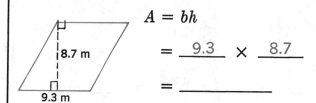

$A = bh$

$= \underline{\quad 9.3 \quad} \times \underline{\quad 8.7 \quad}$

$= \underline{\qquad\qquad}$

The area is _____ square meters.

Find the area of each parallelogram below.

a	b	c

1.

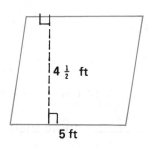

4 ½ ft

5 ft

_____ square feet

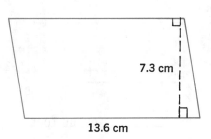

7.3 cm

13.6 cm

_____ square centimeters

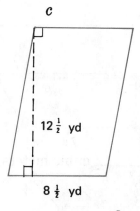

12 ½ yd

8 ½ yd

_____ square yards

Find the area of each parallelogram described below.

	base	height	area
2.	72 mm	24 mm	_____ square millimeters
3.	$7\frac{1}{2}$ ft	5 ft	_____ square feet
4.	$4\frac{3}{4}$ yd	$3\frac{3}{4}$ yd	_____ square yards
5.	7.2 m	6 m	_____ square meters
6.	9.4 cm	6.7 cm	_____ square centimeters
7.	9 yd	$7\frac{1}{4}$ yd	_____ square yards
8.	16 in.	$12\frac{3}{8}$ in.	_____ square inches

Perfect score: 10   My score: _____

Lesson 7 Volume of a Rectangular Solid

The volume measure (V) of a rectangular solid is equal to the product of the area measure of its base (B) and the measure of its height (h). $\qquad V = Bh$

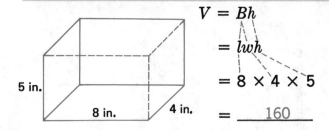

$V = Bh$

$= lwh$

$= 8 \times 4 \times 5$

$= \underline{\quad 160 \quad}$

The volume is _____ cubic inches.

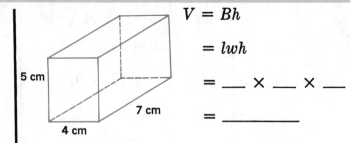

$V = Bh$

$= lwh$

$= \underline{\ } \times \underline{\ } \times \underline{\ }$

$= \underline{\qquad}$

The volume is _____ cubic centimeters.

Find the volume of each rectangular solid below.

a	b	c

1.

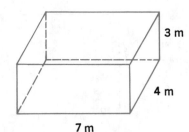

_____ cubic meters

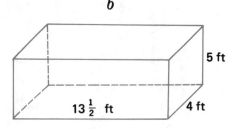

_____ cubic feet

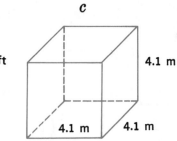

_____ cubic meters

2.

_____ cubic centimeters

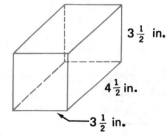

_____ cubic inches

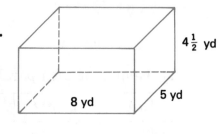

_____ cubic yards

Find the volume of each rectangular solid described below.

	length	width	height	volume
3.	6 cm	7 cm	8 cm	_____ cubic centimeters
4.	4.1 m	3.7 m	2.6 m	_____ cubic meters
5.	$3\frac{1}{2}$ yd	$3\frac{1}{2}$ yd	$3\frac{1}{2}$ yd	_____ cubic yards
6.	28 mm	36 mm	14 mm	_____ cubic millimeters
7.	$7\frac{1}{4}$ ft	$2\frac{1}{2}$ ft	$5\frac{3}{4}$ ft	_____ cubic feet

Perfect score: 11   My score: _____

Lesson 8   Volume of a Cylinder

The volume measure (V) of a cylinder is equal to the product of the area measure of its base (B) and the measure of its height (h). $V = Bh$

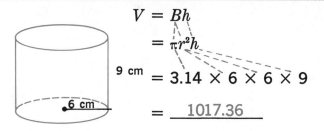

$V = Bh$

$= \pi r^2 h$

$= 3.14 \times 6 \times 6 \times 9$

$= \underline{\quad 1017.36 \quad}$

The volume is _____ cubic centimeters.

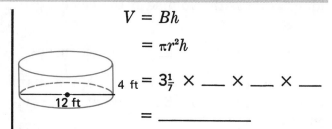

$V = Bh$

$= \pi r^2 h$

$= 3\frac{1}{7} \times \underline{\quad} \times \underline{\quad} \times \underline{\quad}$

$= \underline{\hspace{2cm}}$

The volume is _____ cubic feet.

Find the volume of each cylinder. Use $3\frac{1}{7}$ for π.

a	b	c

1.

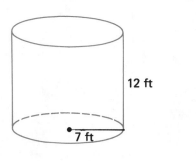

12 ft

7 ft

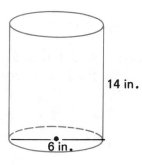

14 in.

6 in.

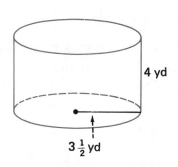

4 yd

$3\frac{1}{2}$ yd

_____ cubic feet   _____ cubic inches   _____ cubic yards

Find the volume of each cylinder described below. Use 3.14 for π.

	radius	height	volume
2.	8 cm	6 cm	_____ cubic centimeters
3.	18 mm	9 mm	_____ cubic millimeters
4.	1.7 m	3.4 m	_____ cubic meters
5.	14 mm	6.5 mm	_____ cubic millimeters
6.	9 cm	14 cm	_____ cubic centimeters
7.	7 m	3.8 m	_____ cubic meters

Perfect score: 9   My score: _____

Problem Solving

Solve each problem. Use 3.14 for π.

1. A box is 6 inches long, 4 inches wide, and 3 inches high. What is the volume of the box?

The volume is _____ cubic inches.

2. A cylindrical storage tank has a diameter of 18 feet and a height of 16 feet. What is the volume of the storage tank?

The volume is _____ cubic feet.

3. Cereal A comes in a rectangular box 20 centimeters wide, 6 centimeters deep, and 25 centimeters high. Find the volume of that box.

The volume is _____ cubic centimeters.

4. Cereal B comes in a cylindrical box that has a diameter of 13 centimeters and a height of 25 centimeters. What is the volume of that box?

The volume is _____ cubic centimeters.

5. Which cereal comes in the box with the larger volume? How much larger?

Cereal _____ came in a box that has a volume

_____ cubic centimeters larger.

6. A classroom is 32 feet long, 24 feet wide, and 10 feet high. What is the volume of the classroom?

The volume is _____ cubic feet.

7. Josephine has a cylindrical juice container with a diameter of 4 inches. Its height is 8 inches. How many cubic inches of juice will the container hold?

The container will hold _____ cubic inches.

1.	2.
3.	4.
5.	6.
7.	

Perfect score: 8   My score: _____

140

Lesson 9   Perimeter, Area, and Volume

Find the perimeter or circumference of each figure below. Use $3\frac{1}{7}$ for π.

a

1.

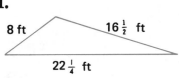

8 ft   16 $\frac{1}{2}$ ft

22 $\frac{1}{4}$ ft

_____ ft

b

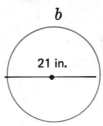

21 in.

_____ in.

Find the area of each figure below. Use 3.14 for π.

a   b   c

2.

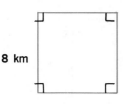

8 km

_____ square kilometers

7 ft

13 ft

_____ square feet

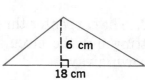

6 cm

18 cm

_____ square centimeters

3.

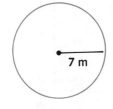

7 m

_____ square meters

9.4 cm

12.7 cm

_____ square centimeters

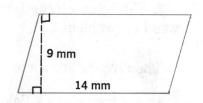

9 mm

14 mm

_____ square millimeters

Find the volume of each figure below. Use $3\frac{1}{7}$ for π.

a   b

4.

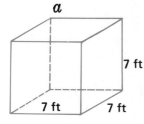

7 ft

7 ft   7 ft

_____ cubic feet

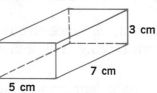

3 cm

7 cm

5 cm

_____ cubic centimeters

5.

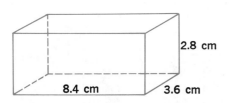

2.8 cm

8.4 cm   3.6 cm

_____ cubic centimeters

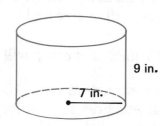

9 in.

7 in.

_____ cubic inches

Perfect score: 12   My score: _____

141

Problem Solving

Solve each problem. Use 3.14 for *pi*.

1. A carpenter cut a circular shelf from a square piece of wood as shown at the right. Find the area of the square piece of wood. Find the area of the circular piece of wood.

80 cm

The area of the square piece is

_____ square centimeters.

The area of the circular piece is

_____ square centimeters.

2. The carpenter threw away the wood left over after cutting out the circular piece. How much wood was thrown away?

_____ square centimeters were thrown away.

3. Find the circumference of the circular piece of wood in problem **1**.

The circumference is _____ centimeters.

4. A farmer has a field shaped like a parallelogram. The base is 1,500 meters. The height is 1,200 meters. Find the area of the field.

The area is _____ square meters.

5. If the farmer puts a fence around the field in problem 4, how much fencing will be needed?

_____ meters of fencing will be needed.

6. How many cubic meters of earth will be removed to dig a well 2 meters in diameter and 28 meters deep?

_____ cubic meters of earth will be removed.

7. A tank is 150 centimeters long, 120 centimeters wide, and 185 centimeters deep. Find its volume.

The volume is _____ cubic centimeters.

1.	
2.	3.
4.	5.
6.	7.

Perfect score: 8   My score: _____

142

CHAPTER 10 TEST

NAME _____

Find the perimeter and the area of each figure below.

| | *a* | *b* | *c* |

1.

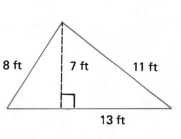

6 m

8 ft   7 ft   11 ft

13 ft

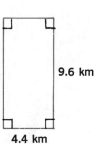

9.6 km

4.4 km

perimeter: _____ meters   _____ feet   _____ kilometers

area: _____ square meters   _____ square feet   _____ square kilometers

2.

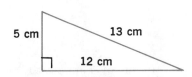

5 cm   13 cm

12 cm

$17\frac{1}{2}$ yd

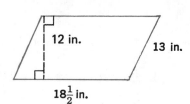

9 yd $13\frac{1}{2}$ yd

12 in.   13 in.

$18\frac{1}{2}$ in.

perimeter: _____ centimeters   _____ yards   _____ inches

area: _____ square centimeters   _____ square yards   _____ square inches

Complete the table below. Use $3\frac{1}{7}$ for π.

	diameter	radius	circumference	area
3.	8 in.	_____ in.	_____ in.	_____ square inches
4.	_____ ft	5 ft	_____ ft	_____ square feet

Find the volume of each figure below. Use 3.14 for π.

| | *a* | *b* |

5.

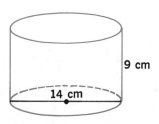

9 cm

14 cm

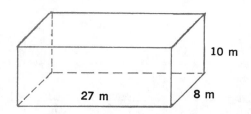

10 m

27 m   8 m

_____ cubic centimeters   _____ cubic meters

Perfect score: 20   My score: _____

PRE-TEST—Graphs

Use the picture graph to help you answer each question.

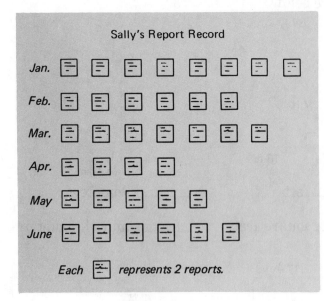

Sally's Report Record

Jan.

Feb.

Mar.

Apr.

May

June

Each ⊟ represents 2 reports.

1. How many reports did Sally write in January? _____ In February? _____

2. In which month did Sally write the greatest number of reports? _____ The least number? _____

3. What was the total number of reports Sally wrote in April, May, and June? _____

Use the bar graph to help you answer each question.

4. In which game was the greatest number of points scored? _____ The least number of points? _____

5. How many points were scored in Game 1? _____ Game 2? _____ Game 3? _____

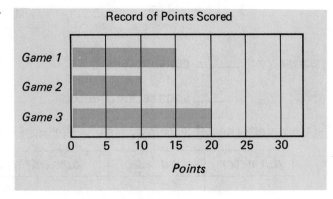

Record of Points Scored

Game 1

Game 2

Game 3

0 5 10 15 20 25 30

Points

Use the line graph to help you answer each question.

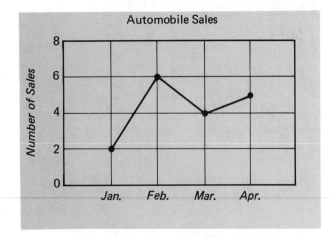

Automobile Sales

Number of Sales

8

6

4

2

0

Jan. Feb. Mar. Apr.

6. How many sales were made in January? _____ In April? _____

7. In which month was the greatest number of sales made? _____ The least number? _____

8. Did the sales increase or decrease from February to March? _____

Perfect score: 15   My score: _____

144

Lesson 1   Picture Graphs

Study how a **picture graph** is used to present the following information in a clear and interesting way.

At Andy's Auto Agency, 20 cars were sold in January, 25 in February, 30 in March, 40 in April, 45 in May, and 50 in June.

Number of Cars Sold (January thru June)

Jan.
Feb.
Mar.
Apr.
May
June

Each ⌐🚗 represents 10 cars.

Since each 🚗 represents 10 cars, each 🚗 must represent _____ cars.

How many cars were sold in March? _____ In February? _____

In which month was the greatest number of cars sold? _____ The least number? _____

Use the picture graphs to help you answer each question.

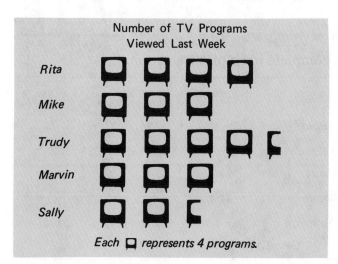

Number of TV Programs
Viewed Last Week

Rita
Mike
Trudy
Marvin
Sally

Each ☐ represents 4 programs.

1. Who watched the greatest number of programs on TV last week? _____ The least number? _____

2. How many programs were watched by Rita? _____ By Sally? _____

3. Which two people watched the same number of programs? _____

4. If each call costs the same amount, which business will have the highest phone bill?

_____ The lowest phone bill? _____

5. On an average day, how many calls were made by Business A? _____ By Business C? _____

6. What is the total number of phone calls made by all the businesses on any one day?

_____

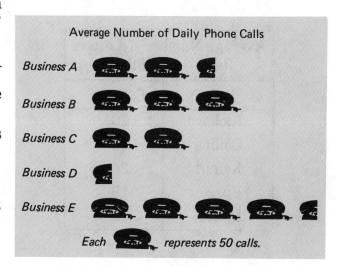

Average Number of Daily Phone Calls

Business A
Business B
Business C
Business D
Business E

Each 📞 represents 50 calls.

Perfect score: 10   My score: _____

145

Lesson 2 Picture Graphs

Andy's Car Sales	
Month	*Cars Sold*
July	30
August	40
September	45
October	30
November	50
December	45

Number of Cars Sold
(July thru December)

July

Aug.

Sept.

Oct.

Nov.

Dec.

Each 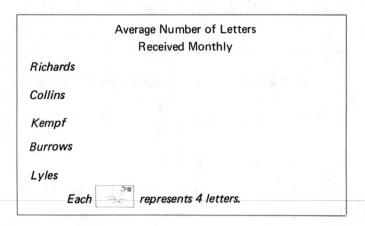 represents 10 cars.

How many 🚗 are drawn for July? _____ August? _____ September? _____

How many 🚗 should you draw for October? _____ November? _____ December? _____

Complete the picture graph.

Use the information in each table to help you complete each picture graph.

1.

Compact Disc Collections	
Student	*Records*
Harold	30
Bill	35
Mary	55
Carol	40
Alice	30

Compact Disc Collections

Harold

Bill

Mary

Carol

Alice

Each ◉ represents 10 discs.

2.

Average Number of Letters Received Monthly	
Family	*Letters*
Richards	20
Collins	30
Kempf	24
Burrows	18
Lyles	22

Average Number of Letters
Received Monthly

Richards

Collins

Kempf

Burrows

Lyles

Each ✉ represents 4 letters.

Perfect score: 10   My score: _____

Lesson 3 Bar Graphs

The length of the first bar on this bar graph shows that Beta Company owns about 250 cars.

How many pickups does

the company own? _____

How many vans does the

company own? _____

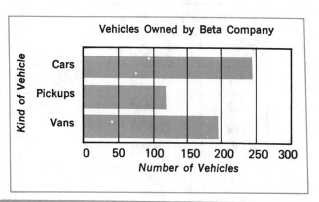

Use the bar graphs to help you complete the following.

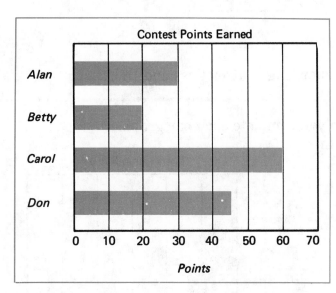

1. Who has the greatest number of points?

_____ The least number? _____

2. How many contest points has Alan

earned? _____ Has Don earned? _____

3. Who has one half as many points as

Carol? _____ One third as many? _____

4. How many points have been earned by

all four persons? _____

5. On which test did Tom receive the highest

score? _____ The lowest score? _____

6. What were Tom's test scores on these

tests: Test 1? ____ Test 3? ____ Test 5? ____

7. On which three tests did Tom improve his

score over the preceding test? _____

8. On which two tests did Tom not improve

his score over the preceding test? _____

9. On which test did Tom receive twice the

score he received on test 1? _____

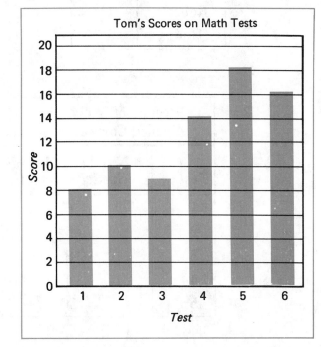

Perfect score: 15   My score: _____

147

Lesson 4   Bar Graphs

Keno County	
Town	*Population*
Ada	3,000
Bay	6,500
Cass	5,000
Dale	2,500

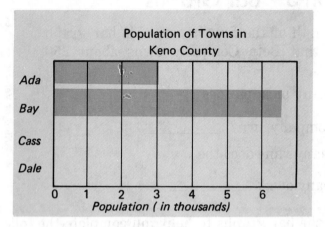

The bar for Ada is drawn to show a population of _____.

The bar for Bay is drawn to show a population of _____.

Complete the graph by drawing bars of proper length after Cass and Dale.

Use the information in each table to help you complete each bar graph.

1.

Membership Record	
Member	*Years*
Ansley	2
Morris	3
Richard	$2\frac{1}{2}$
Roberta	$1\frac{1}{2}$
Sandra	1

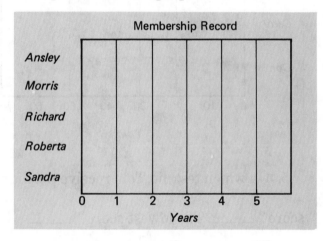

2.

Record of Words Spelled Correctly	
Test	*Number of Words Spelled Correctly*
1	6
2	5
3	8
4	4
5	10

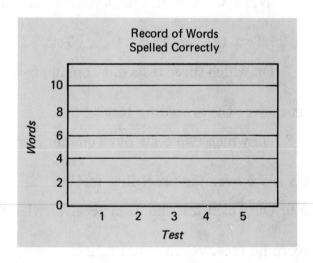

Perfect score: 10   My score: _____

148

Lesson 5   Line Graphs

Study how a line graph is used to show hourly changes in temperature. The temperature at 6 a.m. was 40° and the temperature at 7 a.m. was 45°. What was the temperature at 8 a.m.? _____ At 11 a.m.? _____ Did the temperature increase or decrease from 8 to 9? _____ At what time was the lowest temperature recorded? _____

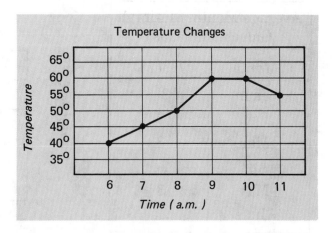

Use the line graphs to help you answer each question.

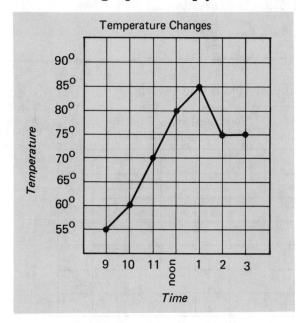

1. What was the temperature at 11 a.m.? _____ At noon? _____ At 1 p.m.? _____

2. What was the lowest temperature recorded? _____ At what time was this? _____

3. What was the highest temperature recorded? _____ At what time was this? _____

4. Did the temperature increase or decrease between 10 and 11? _____ Between 11 and noon? _____ Between 1 and 2? _____

5. In which month was the greatest number of homes sold? _____ The least number? _____

6. In which months was the number of sales the same? _____

7. Did the number of sales increase or decrease during the last 3 months of the year? _____

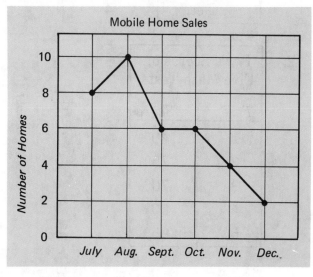

Perfect score: 14   My score: _____

Lesson 6   Line Graphs

Average Monthly Rainfall (Galveston, Texas)	
Month	*Inches*
Jan.	$3\frac{1}{2}$
Feb.	3
Mar.	3
Apr.	$2\frac{1}{2}$
May	3
June	$2\frac{1}{2}$

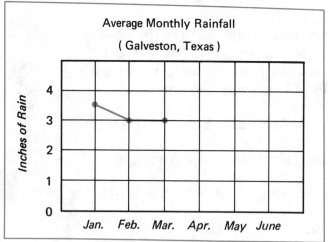

The • for Jan. is drawn to show the average rainfall to be _____ inches.

The • for Feb. represents _____ inches. The • for Mar. represents _____ inches.

Complete the line graph for April, May, and June.

Use the information in each table to help you complete each line graph.

1.

Average Monthly Rainfall (Galveston, Texas)	
Month	*Inches*
July	5
Aug.	$4\frac{1}{2}$
Sept.	5
Oct.	3
Nov.	$3\frac{1}{2}$
Dec.	4

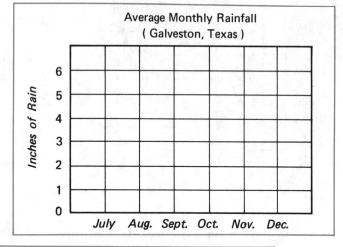

2.

Bev's Scores on Five Math Tests	
Test	*Score*
1	40%
2	75%
3	70%
4	80%
5	95%

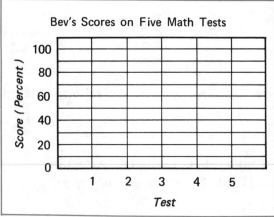

Perfect score: 11   My score: _____

Lesson 7   Circles

A circle and its interior is called a **circular region.**

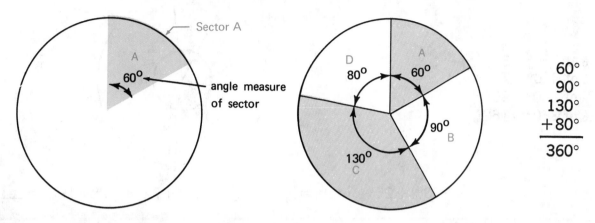

What is the angle measure of sector B? _____ Sector C? _____ Sector D? _____

The sum of the angle measures of all the sectors in a circular region is _____.

Use a protractor to help you separate each circular region as directed. Label each sector with the proper letter and angle measurement.

1. 4 sectors with angle measures as follows:

Sector A	30°
Sector B	60°
Sector C	120°
Sector D	150°

2. 5 sectors with angle measures as follows:

Sector A	20°
Sector B	25°
Sector C	45°
Sector D	90°
Sector E	180°

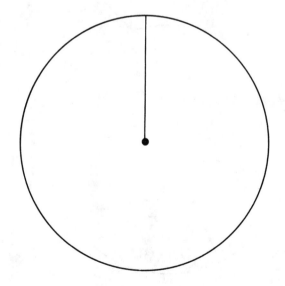

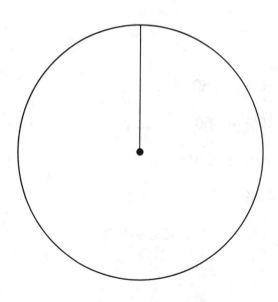

Perfect score: 18   My score: _____

Lesson 8   Circles

Study how the circular region is separated into four sectors representing 10%, 20%, 25%, and 45% of the circular region.

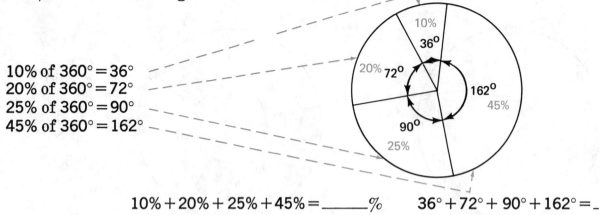

10% of 360° = 36°
20% of 360° = 72°
25% of 360° = 90°
45% of 360° = 162°

10% + 20% + 25% + 45% = _____%   36° + 72° + 90° + 162° = _____°

Complete each sentence. Then write the correct angle measurement in the appropriate sectors.

1.   10% of 360° = _____

    20% of 360° = _____

    30% of 360° = _____

    40% of 360° = _____

    10% + 20% + 30% + 40% = _____

2.   5% of 360° = _____

    15% of 360° = _____

    35% of 360° = _____

    45% of 360° = _____

    5% + 15% + 35% + 45% = _____

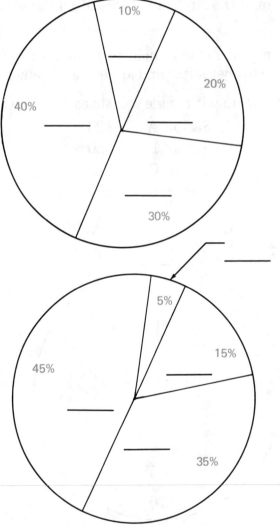

Perfect score: 20   My score: _____

Lesson 9 Circle Graphs

Study how a **circle graph** is used to present the following information in a clear and interesting way.

Arlene spends her allowance as follows: 25% for food, 50% for clothing, 15% for entertainment, and 10% for miscellaneous expenses.

Assume Arlene's allowance is $20.

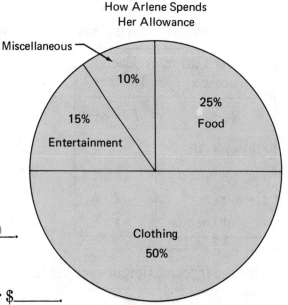

How Arlene Spends
Her Allowance

Miscellaneous
10%
25% Food
15% Entertainment
Clothing 50%

On clothing she would spend 50% of $20 or $__10__.

On food she would spend 25% of $20 or $__5__.

On entertainment she would spend 15% of $20 or $_____.

On miscellaneous expenses she would spend 10% of $20 or $_____.

Complete each sentence.

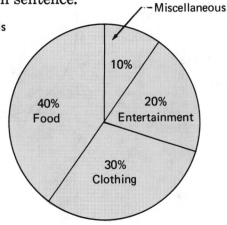

How Lewis Spends
His Allowance

Miscellaneous
10%
40% Food
20% Entertainment
30% Clothing

1. Assume Lewis' allowance is $25.

   He would spend $_____ for clothing.

   He would spend $_____ for food.

   He would spend $_____ for entertainment.

   He would spend $_____ for miscellaneous expenses.

2. Assume Mr. Adams' net income is $9000.

   He would spend $_____ for rent.

   He would spend $_____ for household expenses.

   He would spend $_____ for personal expenses.

   He would save $_____.

   He would spend $_____ for miscellaneous expenses.

How Mr. Adams Spends
His Income

Savings
15% Personal Expenses
5%
20% Rent
50% Household Expenses
Miscellaneous 10%

Perfect score: 9   My score: _____

153

Lesson 10   Circle Graphs

Study how the information in the table can be presented on a circle graph.

Distribution of Each Auto Expense Dollar	
Item	Per cent
Gas and Oil	40%
Depreciation	25%
Repairs	20%
Miscellaneous	15%

40% of 360° = __144°__

25% of 360° = __90°__

20% of 360° = _____

15% of 360° = _____

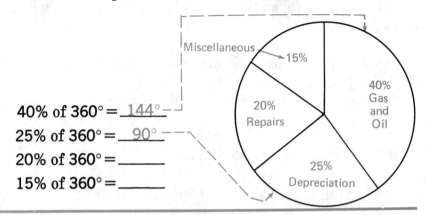

Use the information in each table to help you complete each circle graph.

1.

Distribution of Each Income Dollar	
Expense	Per cent
Rent	30%
Personal	20%
Household	35%
Miscellaneous	15%

30% of 360° = _____

20% of 360° = _____

35% of 360° = _____

15% of 360° = _____

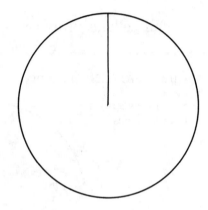

2.

Distribution of Activities on an Average School Day	
Activity	Per cent
Sleeping	30%
School	25%
Eating	10%
Recreation	20%
Miscellaneous	15%

30% of 360° = _____

25% of 360° = _____

10% of 360° = _____

20% of 360° = _____

15% of 360° = _____

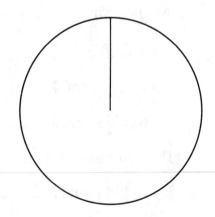

Perfect score: 18   My score: _____

CHAPTER 11 TEST

Use the bar graph to help you answer each question.

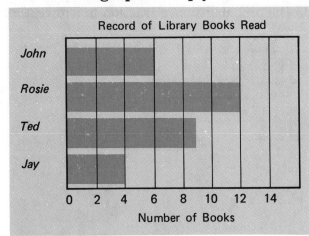

1. Who read the greatest number of library books? _____

2. Who read the least number of library books? _____

3. How many library books were read by John? _____ By Rosie? _____ By Ted? _____

Use the information in the table to help you complete the line graph.

4.

Record of Books Read	
Name	Number
Jane	10
Tom	5
Dick	15
Alice	10
Harry	20

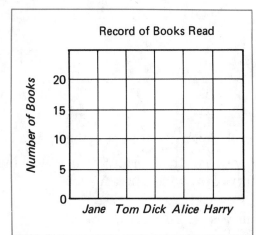

Use the information below to help you complete the sentences and the circle graph.

5.

Distribution of Activities on an Average Work Day	
Activity	Per cent
Work	40%
Travel	5%
Eating	10%
Recreation	15%
Miscellaneous	30%

40% of 360° = _____

5% of 360° = _____

10% of 360° = _____

15% of 360° = _____

30% of 360° = _____

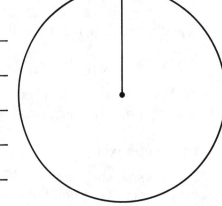

Perfect score: 20   My score: _____

PRE-TEST Probability

You draw one of the cards without looking. Write the probability as a fraction in simplest form that you will draw:

1. Janet _____

2. Jack _____

3. Juan _____

4. a card with a name that starts with a J _____

5. a card with a name that does not start with a J _____

6. Complete the sample space for tossing a penny and a nickel.

penny   *nickel*   *outcome*

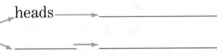
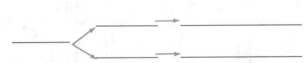

heads

heads _____

_____

_____

_____

_____

_____

Solve each problem. Write each probability as a percent.

7. You draw one of the marbles without looking. What is the probability of drawing a blue marble?

The probability is _____.

8. A company knows that 1% of the bolts they make are defective. If they produce 250,000 bolts, how many will be defective?

_____ bolts will be defective.

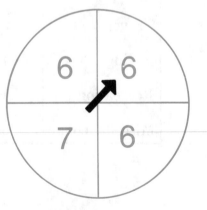

9. You spin the pointer at the right. What is the probability that the pointer will stop on 6?

The probability is _____.

10. You spin the pointer at the right 20 times. Predict how many times the pointer will stop on 6.

The pointer will stop on 6 _____ times.

11. You spin the pointer at the right 200 times. Predict how many times the pointer will stop on 6.

The pointer will stop on 6 _____ times.

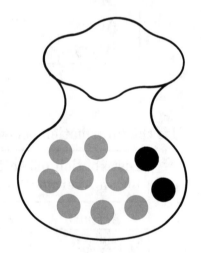

Perfect score: 18   My score: _____

Lesson 1   Probability

You draw one of the cards shown at the right without looking. You would like to know your *chance* or **probability** of getting a card that shows an **A**.

Each card (possible result) is called an **outcome**. There are 6 cards. There are 6 possible outcomes. Since you have the same chance to draw any of the cards, the outcomes are **equally likely**.

A	A	B
B	B	C

number of outcomes
that show **A**

$\dfrac{2}{6}$ or $\dfrac{1}{3}$ Write the probability in simplest form.

number of
possible outcomes

The probability of drawing a card that shows an **A** is $\frac{1}{3}$.

You pick a marble without looking. In simplest form, what is the probability of picking:

1. white _____

2. black _____

3. blue _____

4. a marble that is **not** white _____

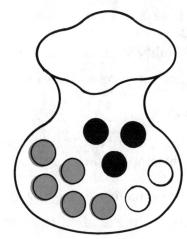

You spin the wheel shown at the right. Find the probability of the spinner stopping on:

5. a blue section _____

6. a white section _____

7. a 1 _____

8. a 3 _____

9. a 2 _____

10. an odd number _____

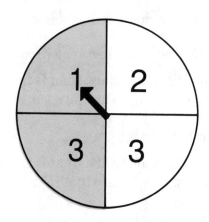

Perfect score: 10   My score: _____

Problem Solving

Your company is having a box-lunch picnic. The company is furnishing free box lunches. The contents are on labels like those at the right. Solve each problem. Write each probability in simplest form.

Cheese Sandwich/ Apple	Roast Beef Sandwich/ Apple
Cheese Sandwich/ Pear	Roast Beef Sandwich/ Orange
Peanut Butter Sand./ Apple	Chicken Sandwich/ Pear
Taco/ Pear	Taco/ Apple

1. Suppose you do not care what kind of lunch you get, so you take one box without looking. What is the probability that you will get a cheese sandwich and an apple?

The probability is _____ .

2. You take one box without looking. What is the probability that you will get an apple?

The probability is _____ .

3. You take one box without looking. What is the probability that you will get a taco?

The probability is _____ .

4. You take one box without looking. What is the probability that you will get a pear?

The probability is _____ .

5. You take one box without looking. What is the probability that you will **not** get an orange?

The probability is _____ .

A game has a board like the one shown below. Use the board to answer each question. Write each probability in simplest form.

6. Are the outcomes equally likely? _____

Draw lines to make all of the rectangles the same size.

DART THROW		
win	lose	win
lose		lose
	win	
win	lose	win
lose		lose
	win	

7. Now how many rectangles say *win*? _____

8. Now how many rectangles say *lose*? _____

9. You throw one dart. What is the probability of hitting a rectangle that says *win*?

The probability is _____ .

10. You throw one dart. What is the probability of **not** hitting a rectangle that says *win*?

The probability is _____ .

Perfect score: 10   My score: _____

Lesson 2   0 and 1 Probabilities

You spin the pointer at the right.

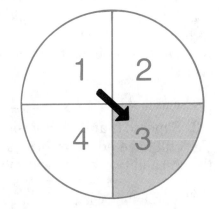

probability of the pointer stopping on 2	probability of the pointer stopping on a number less than 5	probability of the pointer stopping on 6
$\frac{1}{4}$	$\frac{4}{4}$ or 1 A probability of 1 means the outcome is **certain** to happen.	$\frac{0}{4}$ or 0 A probability of 0 means the outcome will **never** happen.

Solve each problem. Write each probability in simplest terms.

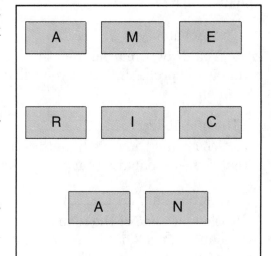

1. You pick one of the letter cards shown at the right without looking. What is the probability that you will pick a vowel (a, e, i, o, u)?

The probability is _____.

2. You pick one of the letter cards without looking. What is the probability that you will pick an A?

The probability is _____.

3. You pick one of the letter cards without looking. What is the probability that you will pick a B?

The probability is _____.

4. You pick one of the letter cards without looking. What is the probability that the letter on the card is in the word *AMERICAN*?

The probability is _____.

5. You pick one of the letter cards without looking. What is the probability that the letter on the card is **not** in the word *AMERICAN*?

The probability is _____.

6. You pick one of the letter cards without looking. What is the probability that the letter on the card is in the word *MINER*?

The probability is _____.

Perfect score: 6   My score: _____

Problem Solving

Solve each problem. Write each probability in simplest form.

1. You are taking a multiple-choice test. Each item has 6 choices. You have no idea which is the correct choice. What is the probability that you will guess the correct choice?

The probability is _____.

2. Suppose that each item on the test in 1 had 4 choices. You still have no idea which is the correct choice. What is the probability that you will guess the correct choice?

The probability is _____.

3. You draw 1 marble from the bag shown at the right. What is the probability that you will draw a marble with a number on it?

The probability is _____.

4. You draw 1 marble from the bag shown at the right. What is the probability that you will draw a marble with a letter on it?

The probability is _____.

5. You draw 1 marble from the bag shown at the right. What is the probability that you will draw a black marble?

The probability is _____.

6. You pick one of the number cards shown at the right without looking. What is the probability that you will pick a number greater than 30?

The probability is _____.

7. You pick one of the number cards shown at the right without looking. What is the probability that you will pick a number less than 100?

The probability is _____.

8. You pick one of the number cards shown without looking. What is the probability that you will pick a card with a 0 on it?

The probability is _____.

9. You pick one of the number cards shown without looking. What is the probability that you will pick a card with a 7 on it?

The probability is _____.

Test   Name _____

In the blank at the left, wri
that best completes the state

_____ 1. The busiest day at the
   a. Monday
   b. Tuesday
   c. Wednesday
   d. Thursday
   e. Friday
   f. Saturday

10	20	30
40	50	60

Perfect score: 9   My score: _____

160

Lesson 3   Sample Spaces

Suppose you flip a penny and a dime. You can show all the possible outcomes in a table or in a tree diagram.

	dime	
	heads	tails
penny	heads	h,h	h,t
	tails	t,h	t,t

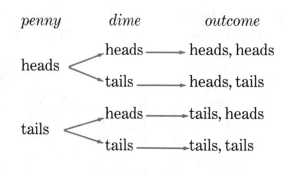

A list or a table of all the possible outcomes is called a **sample space**.

Use the sample spaces above to answer each question.

1. How many possible outcomes are there? _____

2. What is the probability that both coins will land with heads up? _____

3. What is the probability that one coin will land with heads up and the other will land with tails up? _____

Suppose you toss a penny, a nickel, and a dime. Complete the sample space below to show the possible outcomes.

4. penny   nickel   dime   outcome

heads, heads, heads _____

5. How many outcomes are there? _____

6. What is the probability of all heads? _____

7. What is the probability of two heads and one tail? _____

8. What is the probability of at least one tail? _____

Perfect score: 22   My score: _____

Problem Solving

Complete the sample space below to show all the possible outcomes of rolling a blue die and a black die. Then use the sample space to solve each problem. Write each probability in simplest form.

Blue Die

	1	2	3	4	5	6
1	1,1	2,1	3,1	4,1		
2	1,2	2,2	3,2			
3	1,3					
4	1,4					
5						
6						

Black Die

1. What is the probability of rolling two fives?

The probability is _____.

2. What is the probability of rolling the same number on both dice?

The probability is _____.

3. What is the probability of rolling 4, 5 or 5, 4?

The probability is _____.

4. What is the probability of rolling two dice that total 10?

The probability is _____.

5. What is the probability of rolling two dice that total 20?

The probability is _____.

6. What is the probability of rolling two dice that total less than 13?

The probability is _____.

7. What is the probability of rolling two dice that total 7?

The probability is _____.

8. What is the probability of rolling two different numbers?

The probability is _____.

Perfect score: 35   My score: _____

Lesson 4   Probability Experiments

Try this experiment.

Flip a coin 10 times.
How many times did you get heads? _____

Flip a coin 10 more times.
Out of the 20 flips, did you
get heads exactly 10 times? _____

Use tally marks (/) to record the results.

heads	
tails	

Mathematical probability (what you have found in previous lessons) tells what is likely to happen. It does not tell what will actually happen. **Experimental probability** tells what happened during a particular experiment.

Try this experiment. Record your results. Use your results to answer each question.

1. Make paper cards like those shown at the right. Be sure the cards are all the same size. Draw one card without looking, record the result, put the card back. Repeat the experiment 30 times.

Apple	
Orange	
Peach	

Apple	Orange
Apple	Orange
Apple	Peach

2. Based on your experiment, what is the probability of picking a card that says *Apple*? _____

3. What is the mathematical probability of picking a card that says *Apple*? _____

4. Based on your experiment, what is the probability of picking a card that says *Orange*? _____

5. What is the mathematical probability of picking a card that says *Orange*? _____

6. Based on your experiment, what is the probability of picking a card that says *Peach*? _____

7. What is the mathematical probability of picking a card that says *Peach*? _____

8. Repeat the experiment 30 more times. Did your experimental probability results come closer to the mathematical probability after more draws? _____

9. Compare your results with other class members. Did you all get exactly the same results? _____

Perfect score: 9   My score: _____

Problem Solving

Try each experiment. Record your results. Use your results to answer each question.

1. Make paper cards like those shown at the right. Be sure the cards are all the same size. Draw one card without looking, record the result, put the card back. Repeat the experiment 100 times. (You could work with a friend and each person make 50 draws.)

Red	
Blue	
White	
Black	

2. Based on your experiment, what is the probability of picking a card that says *Red*? _____

3. What is the mathematical probability of picking a card that says *Red*? _____

4. Based on your experiment, what is the probability of picking a card that says *Blue*? _____

5. What is the mathematical probability of picking a card that says *Blue*? _____

6. Based on your experiment, what is the probability of picking a card that says *White*? _____

7. Based on your experiment, what is the probability of picking a card that says *Black*? _____

8. Flip a penny and a dime. Record the results with tally marks. Repeat the experiment 25 times.

Penny	*Dime*	
heads	heads	
heads	tails	
tails	heads	
tails	tails	

9. Based on your experiment, what is the probability of both coins landing heads? _____

10. Based on your experiment, what is the probability of at least one coin landing tails? _____

Perfect score: 10   My score: _____

164

Lesson 5 Probability and Percent

You are to draw one marble without looking. The probability of drawing a blue marble is $\frac{5}{10}$ or $\frac{1}{2}$. You can write the probability as a percent in either of these two ways.

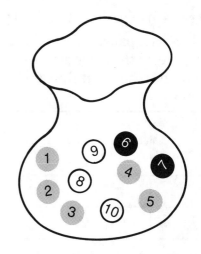

$$\frac{5}{10} = \frac{a}{100}$$

$$500 = 10a$$
$$50 = a$$

$$\frac{5}{10} = \frac{50}{100} = 50\%$$

$$\begin{array}{r} .50 = 50\% \\ 10\overline{)5.00} \\ 5\,0 \\ \hline 00 \\ 00 \\ \hline 0 \end{array}$$

The probability of drawing a blue marble is 50%.

Solve each problem. Write each probability as a percent.

1. Using the bag of marbles at the top of the page, what is the probability of drawing a black marble?

The probability is _____.

2. Using the bag of marbles at the top of the page, what is the probability of drawing a marble with a number on it?

The probability is _____.

You are to choose one of the cards at the right without looking. Write each probability as a percent.

3. What is the probability of choosing *win*?

The probability is _____.

4. What is the probability of not choosing *win*?

The probability is _____.

5. What is the probability of choosing *lose*?

The probability is _____.

6. What is the probability of choosing *draw again*?

The probability is _____.

7. What is the probability of choosing *go home*?

The probability is _____.

win	lose
lose	draw again

Perfect score: 7   My score: _____

165

Problem Solving

Try each experiment. Record your results. Write each probability as a percent.

1. Make paper cards like those shown. Be sure the cards are all the same size. Draw one card without looking, record the result, put the card back. Repeat the experiment 20 times.

Red	
Blue	
White	
Black	

Red	Blue
White	Blue
White	Blue
Black	Black
Black	Black

2. Based on your experiment, what is the probability of picking a card that says *Red*? _____

3. What is the mathematical probability of picking a card that says *Red*? _____

4. Based on your experiment, what is the probability of picking a card that says *Blue*? _____

5. What is the mathematical probability of picking a card that says *Blue*? _____

6. Based on your experiment, what is the probability of picking a card that says *White*? _____

7. What is the mathematical probability of picking a card that says *White*? _____

8. Based on your experiment, what is the probability of picking a card that says *Black*? _____

9. What is the mathematical probability of picking a card that says *Black*? _____

10. What is the experimental probability of picking a card that does **not** say *Black*? _____

11. What is the mathematical probability of picking a card that does **not** say *Black*? _____

12. What is the experimental probability of picking a card that says *Green*? _____

13. What is the mathematical probability of picking a card that says *Green*? _____

Perfect score: 13   My score: _____

Lesson 6   Predicting with Probability

You roll a die once. What is the probability of getting a 5?

Suppose you roll a die 60 times. You can predict how many times you would expect to get a 5 as follows:

probability of getting a 5 — number of rolls

$\frac{1}{6} \times 60 = 10$ number of times you would expect to get a 5

A company finds that 2% of their calculators are defective. Predict how many defective calculators will be defective if they make 5,000 calculators.

$$\begin{array}{r} 5000 \\ \times\ .02 \\ \hline 100.00 \end{array}$$

The company can expect 100 defective calculators.

Solve each problem. Write each probability as a fraction in simplest terms or as a percent.

1. You flip a coin once. What is the probability of getting heads?

The probability is _____.

2. Suppose you flip a coin 200 times. Predict how many times you would expect to get tails.

You should get tails about _____ times.

1–2.

3. A company knows that $\frac{1}{2}$% of the batteries they make are defective. If they produce 100,000 batteries, how many will be defective?

_____ batteries will be defective.

3.

4. You spin the pointer at the right. What is the probability that the pointer will stop on *radio*?

The probability is _____.

5. You spin the pointer at the right 20 times. Predict how many times the pointer will stop on *radio*.

The pointer will stop on *radio* _____ times.

6. You spin the pointer at the right. What is the probability that the pointer will stop on *pencil*?

The probability is _____.

7. You spin the pointer at the right 60 times. Predict how many times the pointer will stop on *pencil*.

The pointer will stop on *pencil* _____ times.

4–7.

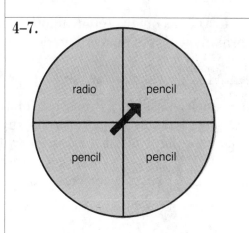

Perfect score: 7   My score: _____

Problem Solving

One hundred people were polled to see whom they preferred for union representative. The results are shown below. Use the results to solve each problem.

Candidate	Erickson	Nunez	Verdugo	Williams
Number of votes	34	27	25	14

1. What percent of those polled prefer Erickson?

_____ % prefer Erickson.

2. If 3,000 people vote for union representative, predict how many will vote for Erickson.

_____ will vote for Erickson.

3. What percent of those polled prefer Williams?

_____ % prefer Williams.

4. If 3,000 people vote for union representative, predict how many will vote for Williams.

_____ will vote for Williams.

5. If 3,000 people vote for union representative, predict how many will vote for Nunez.

_____ will vote for Nunez.

6. If 3,000 people vote for union representative, predict how many will vote for Verdugo.

_____ will vote for Verdugo.

7. Suppose Williams drops out of the election. A poll found that Williams' supporters now support Verdugo. Now predict how many of the 3,000 voters will vote for Verdugo.

_____ will vote for Verdugo.

8. Based on the information in 7, who will win?

_____ will win the election.

9. How many more votes will the winner get than Nunez will get?

The winner will get _____ more votes than Nunez.

1–2.

3–4.

5.

6.

7.

8–9.

Perfect score: 9   My score: _____

Lesson 7   More Probability Experiments

For some events it is hard to find a mathematical probability, so it is necessary to use experimental probability.

Make a cone from a piece of paper like this:

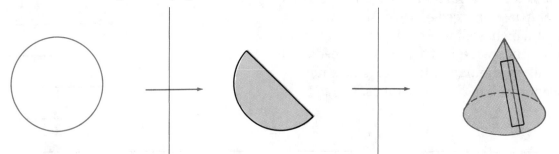

| Trace around a circular object with a diameter between 3 and 6 inches. | Cut out the circle. Fold it in half. Cut along the fold. | Tape the two edges of one semicircle together to make a cone. |

Now toss the cone in the air so it lands on a hard surface. Record the results in the table at the right. Repeat the experiment 50 times.

Landing	Tallies (50 times)
△ (on base)	
△ (on side)	

Use the experiment above. Write each probability as a percent.

1. Based on 50 tries, what is the probability that the cone will land on its base? _____

2. Based on 50 tries, what is the probability that the cone will land on its side? _____

3. Either do the experiment 50 more times or combine your results with those of another student. Record the totals for the 100 tosses in the table at the right.

4. Based on 100 tries, what is the probability that the cone will land on its base? _____

5. Based on 100 tries, what is the probability that the cone will land on its side? _____

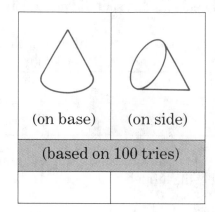

(on base)	(on side)
(based on 100 tries)	

6. Are the results of **1** and **2** close to the results of **3** and **4**? _____

Perfect score: 5   My score: _____

Problem Solving

Do this experiment. Use your results to solve each problem. Write each probability as a percent.

1. Use a sheet of typing paper and a hair pin or needle. On the paper, draw parallel lines so they are just slightly farther apart than the length of the pin or needle. Hold the pin or needle about 6 inches above the paper and let it drop. Record the results. If the pin or needle does not land on the paper, that turn does not count. Repeat the experiment 50 times.

Touched a line	
Did not touch	

2. After 50 tries, what is the probability that the pin or needle landed so it touched a line? _____

3. After 50 tries, what is the probability that the pin or needle landed so it did not touch a line? _____

4. Repeat the experiment 50 more times, or combine your results with those of another student.

Touched a line	
Did not touch	

5. After 100 tries, what is the probability that the pin or needle landed so it touched a line? _____

6. After 100 tries, what is the probability that the pin or needle landed so it did not touch a line? _____

Try the experiment again. This time draw the lines so they are only half as far apart. Record your results for 50 tries below.

Touched a line	
Did not touch	

7. After 50 tries, what is the probability that the pin or needle landed so it touched a line? _____

8. After 50 tries, what is the probability that the pin or needle landed so it did not touch a line? _____

Perfect score: 9   My score: _____

Lesson 8   Problem Solving

Solve each problem. Write each probability as a fraction in simplest form.

1. You choose one of the cards at the right without looking. What is the probability that the card will say *go*?

The probability is _____.

2. You choose one of the cards at the right without looking. What is the probability that the card will say *stop*?

The probability is _____.

3. You choose one of the cards at the right without looking. What is the probability that the card will be *blue*?

The probability is _____.

Solve each problem. Write each probability as a percent.

4. You spin the pointer at the right once. What is the probability that the pointer will stop on 4?

The probability is _____.

5. You spin the pointer 80 times. Predict how many times the pointer would stop on 4.

The probability is _____.

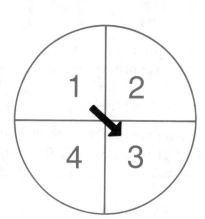

6. A company knows that 2% of their computer disks are defective. The company produced 400,000 computer disks. How many of those disks will be defective?

_____ computer disks will be defective.

Complete this experiment. Flip a coin. Record the results. Repeat the experiment 19 more times. Write each probability as a percent.

7. Based on 20 flips, what is the probability that the coin landed *heads*?

The probability is _____.

8. Based on 20 flips, what is the probability that the coin landed *tails*?

The probability is _____.

9. What is the mathematical probability that a coin will land *heads up*?

The probability is _____.

Perfect score: 9   My score: _____

CHAPTER 12 TEST

You roll a die with faces marked as shown at the right. Write the probability as a fraction in simplest form that you will roll:

ahead 5	back 4	ahead 3
lose a turn	back 1	ahead 2

1. ahead 5 _____

2. back 2 _____

3. back any number _____

4. a face with an odd number _____

5. a face that does not say *back* _____

Solve each problem. Write each probability as a percent.

6. You flip a coin once. What is the probability of getting *tails*?

The probability is _____.

7. Suppose you flip a coin 150 times. Predict how many times you would expect to get tails.

You should get tails about _____ times.

8. A company knows that 3% of the doodads they make are defective. If they produce 300,000 doodads, how many will be defective?

_____ doodads will be defective.

9. You spin the pointer at the right. What is the probability that the pointer will stop on *red*?

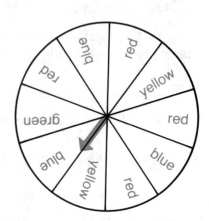

The probability is _____.

10. You spin the pointer at the right 20 times. Predict how many times the pointer will stop on *red*.

The pointer will stop on *red* _____ times.

11. Complete the sample space for tossing a penny and a dime.

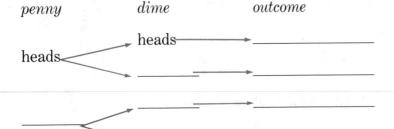

TEST—Chapters 1–5

Complete as indicated. Write each answer in simplest form.

	a	*b*	*c*	*d*	*e*

1.

a
```
   8 3 9 2 0
 +   9 8 2 8
```

b
```
   9 3 2 0 5
 - 2 8 5 9 1
```

c
```
    4 9.2 3 5
 + 3 7.3 6 9
```

d
```
   1.0 3
 - .9 2 5 1
```

e
```
   1 4 5.5
   6 4 3.5 6
       9.2 5
 + 3 2.3
```

2.

a
```
   5 2 4 2
 ×     9 3
```

b
```
   6.2 6
 ×   1 3
```

c
```
 32 ) 3 3 6 0
```

d
```
 .04 ) 5 0.3
```

e
```
   1 2.5 9 3
 ×   .0 3 2
```

3.

a

$$\begin{array}{r} \frac{5}{8} \\ \frac{7}{8} \\ +\frac{5}{8} \\ \hline \end{array}$$

b

$$\begin{array}{r} \frac{3}{4} \\ -\frac{1}{3} \\ \hline \end{array}$$

c

$$\begin{array}{r} 5\frac{1}{6} \\ -2\frac{7}{8} \\ \hline \end{array}$$

d

$$5\frac{1}{4} \times 3\frac{1}{3}$$

e

$$1\frac{3}{5} \div 2\frac{2}{15}$$

Solve each equation.

	a	*b*	*c*

4. $4b = 24$ $\qquad \frac{a}{8} = 13$ $\qquad d + 29 = 120$

5. $h - 5 = 3 \times 7$ $\qquad 6j = 43 + 5$ $\qquad \frac{s}{12} = 5 \times 6$

6. $7a + a = 80$ $\qquad n + n - 1 = 11$ $\qquad r + 3r + 23 = 51$

Continued on the next page.

Test Ch. 1–5

Solve each of the following.

a	b	c
7. $\dfrac{2}{5} = \dfrac{n}{15}$	$\dfrac{9}{n} = \dfrac{1}{4}$	$\dfrac{12}{18} = \dfrac{6}{n}$

Complete each table. Write each fraction in simplest form.

a

	fraction	decimal	percent
8.	$\dfrac{1}{4}$		

b

fraction	decimal	percent
		10%

Complete the following.

a	b	c
9. _____ is 15% of 40	36 is 75% of _____	43 is _____% of 86

Solve each problem.

10. A store sold 185 suits last month. Of those, 60% were women's suits. How many women's suits were sold?

_____ women's suits were sold.

10.

11. Naomi borrowed $2,000 for 1 year at 12% annual interest. How much interest did she pay?

She paid $ _____ interest.

11.

Write an equation for each problem. Solve each problem.

12. Anna made 12 more doodads than Arthur. Together they made 150 doodads. How many doodads did Anna make?

Equation: _____

Anna made _____ doodads.

12.

13. In the drawing at the right, how much weight would have to be applied at point A so the lever would be balanced?

Equation: _____

_____ pounds would have to be applied at point A.

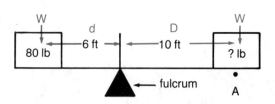

Perfect score: 40   My score: _____

FINAL TEST

NAME _____

Complete as indicated. Write each answer in simplest form.

	a	*b*	*c*	*d*	*e*

1.

$$\begin{array}{r} 1\;5\;3\;2\;3 \\ +\;\;\;5\;6\;2\;8 \\ \hline \end{array}$$

$$\begin{array}{r} 8\;6\;8\;5 \\ -\;8\;5\;9\;1 \\ \hline \end{array}$$

$$\begin{array}{r} 4\;9.0\;5\;3 \\ +\;5\;8.3\;6 \\ \hline \end{array}$$

$$\begin{array}{r} 5\;1.5 \\ -\;\;\;5.5\;7 \\ \hline \end{array}$$

$$\begin{array}{r} 2.5\;3\;5 \\ \times\;.0\;0\;4 \\ \hline \end{array}$$

2.

$$\begin{array}{r} 4\;9\;6 \\ \times\;\;\;5\;4 \\ \hline \end{array}$$

$62\overline{)1\;2\;8\;3\;4}$

$.06\overline{)1.4\;4}$

$$\begin{array}{r} 1\frac{1}{2} \\ +3\frac{1}{2} \\ \hline \end{array}$$

$$\begin{array}{r} 6\frac{1}{3} \\ -4\frac{1}{2} \\ \hline \end{array}$$

3.

$\frac{1}{5} \times 6\frac{2}{3}$

$1\frac{1}{2} \div 2\frac{1}{4}$

$$\begin{array}{r} 6\text{ yd }2\text{ ft} \\ +\;2\text{ yd }2\text{ ft} \\ \hline \end{array}$$

$$\begin{array}{r} 4\text{ pt} \\ -2\text{ pt }1\text{ cup} \\ \hline \end{array}$$

$$\begin{array}{r} 10\text{ min }20\text{ sec} \\ -\;\;7\text{ min }35\text{ sec} \\ \hline \end{array}$$

Solve each of the following.

	a	*b*	*c*

4. $6n = 24$ $\frac{a}{3} = 43$ $e + 9 = 12$

5. $h - 9 = 2 \times 5$ $2j = 23 - 7$ $\frac{s}{8} = 4 \times 3$

6. $2a + a = 30$ $n + n - 5 = 15$ $r + 7r + 14 = 30$

7. $\frac{2}{7} = \frac{n}{14}$ $\frac{9}{n} = \frac{1}{7}$ $\frac{12}{30} = \frac{6}{n}$

Continued on the next page.

Final Test

FINAL TEST (Continued)

Complete each table. Write each fraction in simplest form.

<table>
<tr><td></td><td colspan="3">a</td><td colspan="3">b</td></tr>
<tr><td></td><td>fraction</td><td>decimal</td><td>percent</td><td>fraction</td><td>decimal</td><td>percent</td></tr>
<tr><td>8.</td><td>$\frac{1}{2}$</td><td></td><td></td><td></td><td></td><td>75%</td></tr>
</table>

Complete the following.

	a	b	c
9.	_____ is 25% of 80	105 is 75% of _____	66 is _____ % of 275
10.	6 m = _____ cm	8 km = _____ m	100 liters = _____ kl
11.	100 mg = _____ g	2.3 kg = _____ g	5,600 ml = _____ liters
12.	60 in. = _____ ft	3 pt = _____ cups	4 hours 15 min = _____ min
13.	2 T = _____ lb	32 oz = _____ lb	3 gal 2 qt = _____ qt

Round as indicated.

		nearest thousand	nearest hundred	nearest ten
14.	32,546	_____	_____	_____

Write an estimate for each exercise. Then find the answer.

15.
```
  2,342            40,412             923
 + 923  _____     - 9,342  _____     × 31  _____
```

	a	b
16.	Water freezes at _____ °Celsius or at _____ °Fahrenheit.	

On the _____ before each name below, write the letter of the figure it describes.

17. _____ line segment

18. _____ circle

19. _____ acute angle

20. _____ right triangle

Continued on the next page.

FINAL TEST (Continued)

Use the similar triangles below to help you complete the following.

21. If $a = 16$, $b = 8$, and $d = 12$, then $e =$ _____.

22. If $b = 3$, $c = 5$, and $e = 9$, then $f =$ _____.

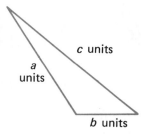

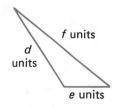

$$\frac{a}{d} = \frac{b}{e} = \frac{c}{f}$$

Find the area of each figure below. Use 3.14 for π.

a	b	c

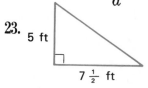

23. 5 ft, $7\frac{1}{2}$ ft

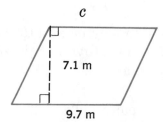 8 m

7.1 m, 9.7 m

_____ square feet   _____ square meters   _____ square meters

Find the volume of each figure below. Use $3\frac{1}{7}$ for π.

24.

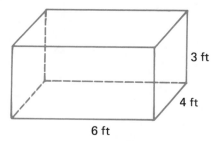

3 ft, 4 ft, 6 ft

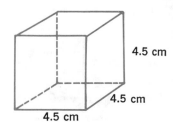

4.5 cm, 4.5 cm, 4.5 cm

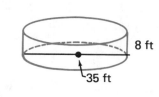

8 ft, 35 ft

_____ cubic feet   _____ cubic centimeters   _____ cubic feet

Use the bar graph to help you answer each question.

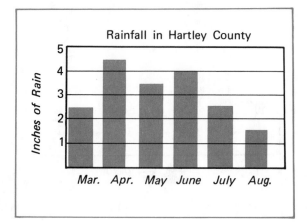

Rainfall in Hartley County

25. In which month was the amount of rain the greatest? _____ least? _____

26. In which months was the amount of rain more than 3 inches? _____

Continued on the next page.

Solve each problem.

27. You draw one of the cards at the right without looking. What is the probability of drawing a card that

says *Win?* _____

says *Lose?* _____

Win	Lose
Lose	Lose
Lose	Lose

28. You make 30 draws, replacing the card after each draw. Predict how many times you will draw a card that

says *Win.* _____ *Lose.* _____

29. A company made 40,000 batteries. It knows 2% will be defective. How many defective batteries did the company make?

The company made _____ defective batteries.

29.

30. Jami borrowed $4,000 for 1 year at 15% annual interest. How much interest did she pay?

She paid $ _____ interest.

30.

31. A sail is shaped as shown. How high is the sail?

x m 5 m 3 m

The sail is _____ meters high.

31.

Write an equation for each problem. Solve each problem.

32. Heather is 2 years older than Jacob. If you add their ages together, the sum is 40. How old is Heather?

Equation: _____

Heather is _____ years old.

32.

33. In the drawing at the right, how much weight would have to be applied at point A so the lever would be balanced?

Equation: _____

_____ pounds would have to be applied at point A.

W 80 lb d 6 ft D 8 ft W ? lb fulcrum A

Perfect score: 80   My score: _____

Answers
Math - Grade 8
(Answers for Pre-Tests and Tests are given on pages 189–191.)

Page 3

	a	b	c	d	e
1.	37	297	5636	47539	653635
2.	76	675	6683	86745	888985
3.	167	898	7917	87864	768854
4.	243	2887	9820	99146	901209
5.	308	3049	24572	127840	1585881

Page 4

	a	b	c	d	e
1.	31	311	4111	14411	301336
2.	45	616	5341	73531	451261
3.	21	809	4625	47329	181109
4.	19	351	3791	11571	345087
5.	38	386	111	64851	353765
6.	18	308	5058	52547	482811
7.	54	413	5867	37835	185989

Page 5

	a	b	c	d	e
1.	252	864	4753	6936	60784
2.	1360	23800	9116	110080	272182
3.	68200	68160	85436	441168	3983250
4.	298200	597550	403300	7484059	6506203

Page 6

	a	b	c	d	e
1.	76 r3	8 r31	12	7 r114	5 r249
2.	203 r4	25 r32	120 r65	14 r1	19 r498
3.	6254	151 r43	2285 r31	102 r16	257 r139

Page 7

1. 806   3. 1400   5. 1825   7. 723 ; 18
2. 221   4. 82   6. 18000

Page 8

1. 419   3. 21587   5. 74375   7. 294
2. 28875   4. 146 ; 4   6. 5922

Page 9

	a	b	c	d	e
1.	3.9	61.4	95.77	35.86	69.967
2.	1.5	23.8	53.48	49.21	75.117
3.	13.9	25.8	97.63	38.74	71.656
4.	33.58	10.658	23.565	17.186	18.593
5.	134.06	4.269	141.061	34.257	178.269

Page 10

1. 8.8   3. 185.45   5. .028   7. 5.84
2. 2.6   4. 10.45   6. 1.009   8. 73.44

Page 11

	a	b	c	d	e
1.	162.5	36.56	268.8	117.46	78.885
2.	2.826	1.561	3.753	.2508	1.3424
3.	.5338	451.58	.29522	.33304	430.94
4.	42.5736	20.7904	320.396	457.678	13.70152
5.	4.6500	1.65164	5.91426	14.3226	9.02275

Page 12

1. 103.8   3. 155.28   5. 30.875   7. 1.075
2. 54   4. 233.75   6. 33.48   8. 7.5

Page 13

	a	b	c	d
1.	5	.5	9	.7
2.	2.7	80	53	.12
3.	26.6	251	20.7	.0367

Page 14

1. 10.2   3. 16   5. 4   7. 255
2. 25   4. 1125   6. 30   8. .03

Page 15

	a	b	c		a	b	c
1.	$\frac{3}{4}$	$2\frac{1}{2}$	$2\frac{3}{5}$	4.	$\frac{1}{2}$	$5\frac{3}{4}$	$1\frac{1}{2}$
2.	$\frac{1}{4}$	$3\frac{3}{5}$	$6\frac{1}{3}$	5.	$\frac{1}{3}$	$2\frac{3}{7}$	$1\frac{3}{5}$
3.	$\frac{5}{8}$	$7\frac{2}{3}$	$8\frac{1}{2}$				

Page 16

	a	b	c	d
1.	3	6	6	10
2.	24	30	21	10
3.	9	27	27	23
4.	24	45	15	8

Page 17

	a	b	c	d
1.	$\frac{4}{5}$	$1\frac{6}{7}$	$\frac{1}{9}$	$1\frac{3}{10}$
2.	$\frac{3}{4}$	$6\frac{1}{6}$	$\frac{1}{2}$	1
3.	$1\frac{5}{12}$	$1\frac{3}{8}$	$11\frac{3}{4}$	$9\frac{9}{20}$
4.	$\frac{1}{6}$	$\frac{1}{5}$	$2\frac{11}{24}$	$5\frac{5}{6}$

Page 18

1. $\frac{3}{4}$ 3. $\frac{3}{4}$ 5. $222\frac{1}{4}$ 7. $1\frac{3}{4}$
2. $\frac{2}{3}$ 4. $6\frac{3}{4}$ 6. $8\frac{1}{2}$ 8. $1\frac{5}{6}$

Page 19

	a	b	c	d
1.	$\frac{3}{20}$	$\frac{8}{15}$	$\frac{1}{30}$	$\frac{9}{40}$
2.	$\frac{3}{25}$	$\frac{7}{12}$	$\frac{5}{32}$	$\frac{1}{3}$
3.	$\frac{2}{3}$	$\frac{1}{4}$	$\frac{1}{2}$	$\frac{7}{32}$
4.	$1\frac{1}{14}$	$\frac{35}{36}$	$\frac{15}{16}$	$1\frac{7}{18}$
5.	2	$2\frac{11}{12}$	$\frac{7}{8}$	$2\frac{11}{12}$
6.	$4\frac{2}{3}$	$17\frac{5}{12}$	$13\frac{1}{8}$	$24\frac{4}{15}$

Page 20

1. $\frac{3}{8}$ 3. $4\frac{3}{8}$ 5. $\frac{5}{8}$ 7. $12\frac{1}{2}$
2. $\frac{5}{6}$ 4. $2\frac{1}{2}$ 6. $10\frac{7}{8}$ 8. 39

Page 21

	a	b	c	d
1.	$\frac{2}{3}$	$1\frac{5}{16}$	$1\frac{2}{5}$	$\frac{25}{28}$
2.	$\frac{1}{5}$	12	$\frac{3}{10}$	15
3.	$2\frac{1}{4}$	4	$3\frac{4}{7}$	$4\frac{20}{21}$
4.	$\frac{7}{20}$	$\frac{1}{2}$	$\frac{5}{16}$	$\frac{5}{12}$
5.	$1\frac{1}{15}$	$1\frac{1}{2}$	$5\frac{1}{3}$	$\frac{2}{3}$
6.	$\frac{9}{16}$	$1\frac{11}{15}$	3	$4\frac{2}{3}$

Page 22

1. $\frac{5}{8}$ 3. 8   5. $\frac{3}{8}$ 7. 48
2. $\frac{3}{16}$ 4. $1\frac{1}{8}$ 6. $3\frac{1}{3}$ 8. 12

Page 25

	a	b
1.	$11-c$	$b+5$
2.	$d+12$	$\frac{t}{2}$ or $t \div 2$
3.	$8 \times n$ or $8n$	$x-4$
4.	$\frac{8}{j}$ or $8 \div j$	$\frac{1}{2} \times y$ or $\frac{1}{2}y$
5.	$12-3$ or 9	
6.	$7+9$ or 16	
7.	$48 \div 3$ or 16	
8.	$\frac{1}{4} \times 72$ or 18	

9. 4×12 or 48
10. $6-6$ or 0
11. $24 \div 3$ or 8

Page 26

	a	b
1.	$a+6=20$	$p-7=15$
2.	$20 \div y = 4$ or $\frac{20}{y}=4$	$\frac{1}{2}t=14$
3.	$b+7=14$	$v+12=18$
4.	$2n=12$	$d \div 3 = 14$ or $\frac{d}{3}=14$

5. 4 ; 4   7. 12 ; 12   9. 6 ; 6   11. 8 ; 8
6. 5 ; 5   8. 14 ; 14   10. 2 ; 2

Page 27

	a	b	c
1.	$w=4$	$b=17$	$m=12$
2.	$a=36$	$c=18$	$r=19$
3.	$e=14$	$s=3$	$d=17$
4.	$x=3$	$m=5$	$y=24$

Page 28

1. $9x$; $9x=36$; 4 ; 4
2. $70x$; $70x=630$; 9 ; 9
3. $3x$; $3x=234$; 78 ; 78
4. $8x$; $8x=784$; 98 ; 98
5. $12x$; $12x=84$; 7 ; 7

Page 29

	a	b	c
1.	$a=56$	$b=117$	$c=72$
2.	$r=128$	$s=56$	$t=90$
3.	$g=85$	$h=75$	$j=126$
4.	$m=255$	$n=644$	$p=513$

Page 30

2. $\frac{x}{3} \cdot \frac{x}{3} = 48$; 144 ; 144
3. $\frac{x}{6} \cdot \frac{x}{6} = 3$; 18 ; 18
4. $\frac{x}{3} \cdot \frac{x}{3} = 120$; 360 ; 360
5. $\frac{x}{5} = 12$; 60 ; 60

Page 31

	a	b	c
1.	$d=36$	$e=48$	$f=22$
2.	$j=25$	$h=18$	$l=27$
3.	$w=64$	$x=76$	$y=220$
4.	$a=38$	$b=146$	$c=55$

Page 32

2. $x+27$; $x+27=45$; 18 ; 18
3. $x+8$; $x+8=94$; 86 ; 86
4. $x+5$; $x+5=17$; 12 ; 12
5. $x+19$; $x+19=48$; 29 ; 29

Page 33

	a	b	c
1.	$b=23$	$x=50$	$c=35$
2.	$r=54$	$e=106$	$f=45$
3.	$a=190$	$m=119$	$t=130$
4.	$w=200$	$g=258$	$h=231$

Page 34

2. $x-7$; $x-7=78$; 85 ; 85
3. $x-324$; $x-324=126$; 450 ; 450
4. $x-95$; $x-95=1550$; 1645 ; 16.45
5. $x-37$; $x-37=75$; 112 ; 112

Page 35

	a	b	c
1.	$b=15$	$u=13$	$v=12$
2.	$d=300$	$y=224$	$k=74$
3.	$g=18$	$x=54$	$n=31$
4.	$p=19$	$k=95$	$w=132$
5.	$x=29$	$m=140$	$n=52$
6.	$t=14$	$d=135$	$n=60$
7.	$q=144$	$g=117$	$x=225$

Page 36

1. $3x=48$; 16 ; 16
2. $\frac{1}{3}x=48$; 144 ; 144
3. $x-27=48$; 75 ; 75
4. $x+27=48$; 21 ; 21

Page 39

	a	b	c		a	b
1.		$7e$	$9f$	7.		9
2.	g	$4h$	$4j$	8.	8	2
3.	$3k$	$2l$	$5m$	9.	20	4
4.	$8n$	p	$3q$	10.	8	5
5.	$6r$	$9s$	$6t$	11.	18	3
6.	$7u$	$8v$	$4w$	12.	30	5

Page 40

	a	b	c
1.	$a=5$	$b=9$	$c=7$
2.	$d=10$	$e=10$	$f=18$
3.	$g=12$	$h=15$	$j=9$
4.	$k=20$	$l=18$	$m=13$
5.	$n=12$	$p=24$	$q=24$

Page 41

1. $x+3x=28$; 7 ; 21
2. $x+4x=75$; 15 ; 60
3. $x+6x=4200$; 600 ; 3600
4. $x+3x=52$; 13 ; 39

Page 42

1. $x+(x+7)=55$; 31 ; 24
2. $x+(x+6)=58$; 26 ; 32
3. $x+(x+16)=184$; 84 ; 100
4. $x+(x+12)=150$; 69 ; 81

Page 43

1. $x+(2x-40)=170$; 70 ; 100
2. $x+(3x+3)=43$; 10 ; 33
3. $x+(3x+2)=10$; 2 ; 8

Page 44

1. $x+(x+2)=16$; 9 ; 7
2. $x+2x=27$; 9 ; 18
3. $x+(2x+1)=22$; 15 ; 7
4. $x+(x+1)=97$; 48 ; 49

Page 45

1. $d=450 \times 5$; 2250
2. $744 = r \times 12$; 62
3. $d=25 \times 48$; 1200
4. $780 = r \times 2$; 390
5. $d=204.8 \times 4$; 819.2

Page 46

1. $60 \times 2 = 40 \times D$; 3
2. $600 \times 12 = W \times 36$; 200
3. $240 \times .7 = W \times .5$; 336

180

Page 49

	a	b		a	b
2.	2 to 4	$\frac{2}{4}$	8.	3 to 72	$\frac{3}{72}$
3.	4 to 3	$\frac{4}{3}$	9.	5 to 20	$\frac{5}{20}$
4.	3 to 3	$\frac{3}{3}$	10.	5 to 4	$\frac{5}{4}$
5.	5 to 4	$\frac{5}{4}$	11.	11 to 2	$\frac{11}{2}$
7.	5 to 6	$\frac{5}{6}$	12.	6 to 9	$\frac{6}{9}$

Page 50

	a	b
1.	$\frac{2}{3}=\frac{8}{12}$	
2.		$\frac{5}{6}=\frac{20}{24}$
3.	$\frac{1}{6}=\frac{2}{12}$	$\frac{14}{16}=\frac{7}{8}$
4.	$\frac{5}{12}=\frac{15}{36}$	
5.		
6.	$\frac{1}{3}=\frac{6}{18}$	$\frac{15}{24}=\frac{5}{8}$
7.		
8.	$\frac{9}{10}=\frac{90}{100}$	$\frac{8}{10}=\frac{4}{5}$
9.		
10.	$\frac{12}{25}=\frac{48}{100}$	$\frac{125}{1000}=\frac{1}{8}$

Page 51

	a	b	c
1.	$n=12$	$n=15$	$n=75$
2.	$n=8$	$n=12$	$n=48$
3.	$n=4$	$n=7$	$n=2$
4.	$n=16$	$n=7$	$n=5$
5.	$n=3$	$n=625$	$n=48$

Page 52

1. 90		3. 600		5. 4
2. 49		4. 15		

Page 53

1. 3	3. 42	5. 42	7. 57
2. 60	4. 24	6. 120	

Page 54

1. 2	3. 4200	5. 63	7. 30
2. 2800	4. 8750	6. 45	

Page 55

	a	b		a	b
1.	$\frac{3}{100}$	.03	8.	$\frac{11}{100}$	.11
2.	$\frac{27}{100}$	.27	9.	$\frac{167}{100}$	1.67
3.	$\frac{121}{100}$	1.21	10.	$\frac{57}{100}$	.57
4.	$\frac{7}{100}$	.07	11.	$\frac{251}{100}$	2.51
5.	$\frac{39}{100}$	.39	12.	$\frac{69}{100}$	.69
6.	$\frac{141}{100}$	1.41	13.	$\frac{391}{100}$	3.91
7.	$\frac{9}{100}$	.09	14.	$\frac{87}{100}$	.87

Page 56

	a	b		a	b
1.	25	$37\frac{1}{2}$	6.	$\frac{1}{10}$	$\frac{4}{5}$
2.	10	75	7.	$1\frac{3}{5}$	$\frac{1}{8}$
3.	50	$62\frac{1}{2}$	8.	$2\frac{1}{2}$	$\frac{5}{8}$
4.	70	40	9.	$\frac{1}{5}$	$\frac{4}{25}$
5.	80	$87\frac{1}{2}$	10.	$1\frac{1}{4}$	$\frac{3}{8}$

Page 57

	a	b		a	b
1.	20%	.52	5.	147%	1.8374
2.	190%	1.48	6.	6.7%	.095
3.	2%	.054	7.	12.3%	.0725
4.	36%	.0875	8.	162.5%	.0875

Page 58

	a	b		a	b
1.	$12\frac{1}{2}$	$\frac{3}{10}$	9.	50	.04
2.	$\frac{4}{5}$	20	10.	.177	110
3.	60	$1\frac{1}{5}$	11.	67	.06625
4.	$\frac{7}{8}$	75	12.	.0846	158
5.	10	$1\frac{1}{2}$	13.	12.5	.04075
6.	$\frac{5}{16}$	90	14.	.06007	31.2
7.	16	$\frac{16}{25}$	15.	.0625	.0975
8.	$1\frac{1}{10}$	45	16.	.0775	.055

Page 59

	a	b		a	b
1.	8	4.8	6.	408	5.9
2.	96	29.76	7.	24	70
3.	22.77	57	8.	6.72	46.08
4.	525	5.561	9.	432	6.225
5.	$\frac{3}{16}$	5.07	10.	14.4	124.8

Page 60

1. 646	3. 27	5. 420	7. 2.34
2. 36	4. 200	6. $1\frac{7}{8}$	

Page 61

	a	b		a	b
1.	50	80	6.	80	25
2.	110	100	7.	75	20
3.	50	25	8.	150	60
4.	25	80	9.	25	80
5.	$62\frac{1}{2}$	25	10.	$12\frac{1}{2}$	25

Page 62

1. $62\frac{1}{2}$	3. 20	5. $87\frac{1}{2}$	7. 150
2. 140	4. 85	6. 70	

Page 63

	a	b		a	b
1.	185	184	6.	24	5.2
2.	13.6	40	7.	29	90
3.	100	30	8.	9	2
4.	125	130	9.	600	60
5.	4400	700	10.	256	200

Page 64

1. 56	3. 300	5. 400	7. 300
2. 700	4. 480	6. 880	

Page 65

	a	b		a	b
1.	12	20	9.	48.96	50
2.	52	40	10.	51.2	986
3.	75	19	11.	140	$65\frac{1}{3}$
4.	74	216	12.	864	220
5.	72	25	13.	40	40
6.	23.04	50	14.	600	100
7.	50	49.2	15.	30	$1\frac{1}{2}$
8.	5.2	36	16.	1	292

Page 66

1.	306	3.	75	5.	111	7.	429
2.	95	4.	24	6.	300		

Page 69

1.	$50	6.	$56.25
2.	$96	7.	$300
3.	$55	8.	$375
4.	$224.25	9.	$1011.50
5.	$288	10.	$585

Page 70

1.	81	3.	38.50	5.	102
2.	8.75	4.	1200 ; 11200	6.	23 ; 34.50

Page 71

	p	r	t	i
1.	$100			
2.		8%		
3.			$\frac{1}{2}$ year	
4.				$234
5.			2 years	
6.		8%		
7.	$3500			
8.		$7\frac{1}{2}\%$		
9.			3 years	
10.				$11250

Page 72

1.	400	3.	$\frac{1}{2}$	5.	600	7.	3400
2.	12	4.	9.25 ; 749.25	6.	15		

Page 73

	total amount		total amount
1.	$561.80	4.	$966.36
2.	$779.12	5.	$259.01
3.	$926.10	6.	$1166.40

Page 74

1.	674.16	3.	367.51 ; 363	5.	Bill ; 115.70
2.	441 ; 463.05	4.	578.81		

Page 75

	total amount		total amount
1.	$218.55	4.	$421.37
2.	$331.15	5.	$513.47
3.	$105.10	6.	$613.60

Page 76

1.	646.14	3.	106 ; 106.09 ; 106.14	5.	A ; .41
2.	442.58	4.	1.62		

Page 79

	a	b
1.	liters	grams
2.	meters	meters
3.	liters	grams
4.		1000 grams
5.	.01 gram	.01 meter
6.	.001 liter	.001 meter
7–9.	Answers will vary.	

Page 80

	a	b	c		a	b	c
1.	1 km	1 dm	1 km	4.	1	1	1
2.	1 dm	1 km	1 cm	5.	1	.001	1
3.	100	1000	10				

Page 81

	a	b		a	b
1.	5000	452	6.	9.2	.7
2.	.038	.948	7.	500	3
3.	750	8000	8.	.092	3.6
4.	.04	.75	9.	.2 km	
5.	920	48.6	10.	1000 m ; 1 km	

Page 82

	a	b
1.	10 kiloliters	1 kiloliter
2.	1000 liters	1 liter
3.	1000 liters	.001 kiloliter
4.	1 kiloliter	1 liter

	a	b		a	b
5.	1000	100	7.	1	1
6.	1000	10	8.	.1	10000

Page 83

	a	b		a	b
1.	6400	6	6.	7500	:0075
2.	25000	.752	7.	less than	
3.	78000	.529	8.	10	
4.	986	.042	9.	15 ; .015	
5.	7500	.0075			

Page 84

	a	b		a	b
1.	26000	6200	5.	60500	.0605
2.	.0752	2.42	6.	5 ; 5000	
3.	89000	7500	7.	1000 ; 1	
4.	.835	.0056			

Page 85

1.	950	3.	45000	5.	4.20	7.	72
2.	Rosa ; 2	4.	250	6.	A ; .3		

Page 86

1.	32	8.	10° Fahrenheit
2.	0	9.	86° Fahrenheit
3.	212	10.	30° Celsius
4.	100	11.	28° Fahrenheit
5.	37	12.	112° Celsius
6.	98.6	13–14.	Answers will vary.
7.	28° Celsius		

Page 89

	a	b		a	b
1.	4	64	7.	9	221
2.	84	19	8.	$3\frac{1}{2}$	11
3.	14	81	9.	45	$105\frac{1}{2}$
4.	15	153	10.	$2\frac{1}{4}$	$27\frac{1}{2}$
5.	2	17	11.	198	$41\frac{1}{4}$
6.	252	55			

Page 90

1.	5	3.	72	5.	60	7.	3	9.	84
2.	74	4.	160	6.	60	8.	66		

Page 91

	a	b		a	b
1.	330	134	7.	$1\frac{1}{5}$	231
2.	3	220	8.	8	166
3.	$3\frac{3}{4}$	23	9.	4	316
4.	210	7	10.	64	135
5.	6000	5	11.	72	14
6.	3	101			

Page 92

1. 205	3. 16	5. 2	7. 103
2. 9	4. 40	6. 17	8. 4000

Page 93

	a	*b*		*a*	*b*
1.	3	13	2.	1	11

	a	*b*		*a*	*b*
3.	19	2	4.	1	1

	a	*b*
5.	9 yd 2 ft	5 min 43 sec
6.	5 lb 11 oz	10 ft 10 in.
7.	6 yd 1 ft	13 lb 6 oz
8.	10 lb 1 oz	7 gal 1 qt
9.	12 ft	9 lb

	c	*d*
5.	5 gal 3 qt	9 ft 7 in.
6.	5 hours 37 min	9 pt
7.	12 min 13 sec	8 hours 17 min
8.	6 min 10 sec	6 lb 4 oz
9.	6 pt	13 yd

Page 94

1. 3 ; 21	3. 12 ; 3	5. 11 ; 5	7. 9 ; 28
2. 10 ; 2	4. 6 ; 2	6. 6 ; 5	

Page 95

	a	*b*		*a*	*b*
1.	16	19	3.	3	105
2.	6	5	4.	63	3

	a	*b*
5.	6 ft 1 in.	4 lb 5 oz
6.	7 ft	5 yd 1 ft
7.	2 lb 14 oz	1 pt
8.	1 gal 3 qt	2 yd 2 ft
9.	3 min 9 sec	3 gal 3 qt

	c	*d*
5.	1 hour 26 min	4 gal 1 qt
6.	3 qt	3 min 11 sec
7.	1 hour 34 min	5 ft 7 in.
8.	7 lb 10 oz	6 min 58 sec
9.	3 lb 5 oz	4 pt 1 cup

Page 96

1. 8 ; 12	3. 5	5. 2 ; 3	7. 1 ; 2
2. 31	4. 3 ; 1	6. 45	

Page 97

	a	*b*	*c*
1.	16 ft 8 in.	12 min 48 sec	9 gal 3 qt
2.	10 hours	25 lb 10 oz	14 yd 2 ft
3.	46 ft 6 in.	45 gal 2 qt	31 lb 8 oz
4.	65 lb 13 oz	39 ft 6 in.	48 min
5.	29 yd 1 ft	32 min	10 lb 11 oz
6.	10 lb 15 oz	7 hours	19 qt 1 pt
7.	42 yd 2 ft	28 ft	64 lb 11 oz

Page 98

1. 25 ; 6	3. 21 ; 6	5. 66 ; 8	7. 67 ; 30
2. 6 ; 45	4. 12 ; 1	6. 85 ; 8	

Page 99

	a	*b*
1.	120	43
2.	80	$43\frac{1}{2}$
3.	$1\frac{3}{4}$	218

Page 99 (continued)

4.	6	9
5.	4	17

	a	*b*	*c*
6.	8 ft 11 in.	2 lb 7 oz	22 qt 1 pt
7.	21 hours	31 ft 8 in.	3 gal 1 qt
8.	2 ; 9	9. 122	

Page 100

	a	*b*	*c*		*a*	*b*	*c*
1.	30	70	90	6.	24100	35300	47400
2.	240	480	660	7.	4000	7000	8000
3.	1700	2790	8250	8.	6000	8000	9000
4.	300	500	600	9.	16000	28000	44000
5.	1500	2600	1700				

Page 101

	a		*b*		*c*	
1.	1174	1100	433	500	1568	1500
2.	5928	6000	7802	7000	1549	2000
3.	11066	12000	2326	2000	11513	12000
4.	12568	20000	99567	100000	36621	40000

Page 102

	a		*b*		*c*	
1.	2736	2800	5187	5400	3575	4200
2.	3312	3500	4368	4800	4950	5600
3.	5292	4800	3999	3600	3330	3500
4.	9750	8000	16884	20000	45612	42000
5.	18361	16000	60225	56000	45778	50000

Page 105

4. line RS or SR ; None ; $\overleftrightarrow{RS}$ or $\overleftrightarrow{SR}$

5. line segment XY or YX ; X and Y ; $\overline{XY}$ or $\overline{YX}$

6. ray TV ; T ; $\overrightarrow{TV}$

7. ray CA ; C ; $\overrightarrow{CA}$

8. line WZ or ZW ; None ; $\overleftrightarrow{WZ}$ or $\overleftrightarrow{ZW}$

9. line segment HN or NH ; H and N ; $\overline{HN}$ or $\overline{NH}$

Page 106

	c	*r*	*d*
1.	point B	$\overline{BD}, \overline{BA},$ or $\overline{BC}$	$\overline{AC}$ or $\overline{CA}$
2.	point K	$\overline{KJ}$ or $\overline{KL}$	$\overline{JL}$ or $\overline{LJ}$
3.	True	4. True	5. True

Page 107

	a	*b*	*c*
1.	obtuse	right	acute
2.	acute	obtuse	right

Page 108

	a	*b*	*c*		*a*	*b*	*c*
1.	90	60	125	2.	45	100	75

Page 109

	a	*b*		*a*	*b*
1.	30°	90°	4.	30°	120°
2.	50°	70°	5.	80°	150°
3.	40°	120°			

6–7. Have your teacher check your work.

Page 110

	a	*b*
1.	congruent	congruent
2.	not congruent	congruent
3.	25	150

Page 111

	a	b	c
1.	perpendicular	parallel	neither
2.	parallel	parallel	perpendicular
3.	parallel	neither	perpendicular

Page 112

	a	b	c
1.	obtuse	right	acute
2.	scalene	isosceles	equilateral or isosceles

Page 115

	a	b	
1.	D'E'F'	JKL	3. False
2.	P'	M'	4. True
	Q'	N'	5. True
	R'	O'	6. False

Page 116

1. $\frac{3}{6}$ $\frac{1}{2}$ $\frac{5}{10}$ $\frac{1}{2}$ $\frac{7}{14}$ $\frac{1}{2}$
2. $\frac{8}{6}$ $\frac{4}{3}$ $\frac{12}{9}$ $\frac{4}{3}$ $\frac{16}{12}$ $\frac{4}{3}$
3. $\frac{12}{18}$ $\frac{2}{3}$ $\frac{8}{12}$ $\frac{2}{3}$ $\frac{10}{15}$ $\frac{2}{3}$

Page 117

	a	b		a	b		a	b
1.	10	8	2.	24	50	3.	12	12

Page 118

1. 32   2. 20   3. 25   4. 40

Page 119

	a	b
1.	$5 \times 5 = 25$	$\sqrt{5 \times 5} = 5$
2.	$8 \times 8 = 64$	$\sqrt{8 \times 8} = 8$
3.	$2 \times 2 = 4$	$\sqrt{2 \times 2} = 2$
4.	$10 \times 10 = 100$	$\sqrt{10 \times 10} = 10$
5.	$4 \times 4 = 16$	$\sqrt{4 \times 4} = 4$
6.	$12 \times 12 = 144$	$\sqrt{12 \times 12} = 12$
7.	$20 \times 20 = 400$	$\sqrt{20 \times 20} = 20$
8.	$11 \times 11 = 121$	$\sqrt{11 \times 11} = 11$
9.	$19 \times 19 = 361$	$\sqrt{19 \times 19} = 19$
10.	$25 \times 25 = 625$	$\sqrt{25 \times 25} = 25$
11.	$31 \times 31 = 961$	$\sqrt{31 \times 31} = 31$
12.	$43 \times 43 = 1849$	$\sqrt{43 \times 43} = 43$
13.	$50 \times 50 = 2500$	$\sqrt{50 \times 50} = 50$

Page 120

1. 324 4.24
2. 625 5
3. 2025 6.71
4. 4096 8
5. 6889 9.11
6. 5625 8.66
7. 8100 9.49
8. 10816 10.2
9. 18225 11.62
10. 21609 12.12
11. 22500 12.25

Page 122

	a	b	c		a	b	c
1.	13	23	28	7.	101	105	107
2.	16	19	31	8.	112	132	149
3.	35	47	63	9.	70	43	130
4.	41	56	69	10.	100	110	50
5.	73	78	83	11.	92	139	118
6.	89	93	98	12.	91	135	146

Page 123

1. 10   5. 13   9. 29
2. 25   6. 11.31   10. 12.17
3. 8.6   7. 17   11. 53
4. 11.4   8. 8.54

Page 124

1. 13   2. 10   3. 8.6   4. 65   5. 12.08

Page 125

1. 7   5. 10   9. 11.96
2. 9   6. 10.54   10. 55
3. 11   7. 39   11. 12.17
4. 11.87   8. 16   12. 77

Page 126

1. 11.31   3. 30   5. 11.53
2. 7   4. 65   6. 21

Page 127

1. 5 ; 24 ; 26
2. 4 ; 4.5 ; 7.5
3. 8 ; $18\frac{3}{4}$; $21\frac{1}{4}$

Page 128

1. 3 ; 8 ; 10
2. 29 ; $43\frac{1}{2}$; $31\frac{1}{2}$
3. 24 ; 5 ; 10

Page 131

	a	b	c
1.	$23\frac{1}{2}$	40	29.6
2.	$33\frac{1}{3}$	21.8	42
3.	$31\frac{1}{2}$	41.6	42

Page 132

	a	b	c
1.	44	$17\frac{3}{5}$	33
2.	18.84	131.88	
3.	47.1	42.076	
4.	21.352	301.44	
5.	254.34	232.36	
6.	84.78	60.288	
7.	13.188	25.12	

Page 133

1. 91 ; $72\frac{1}{4}$; 61.75
2. 891
3. $18\frac{3}{8}$
4. $7\frac{1}{2}$
5. 44.89
6. 70.84
7. 82.8
8. 12.96
9. 62.7

Page 134

1. 14 ; $13\frac{1}{8}$; 20
2. $67\frac{1}{2}$
3. $11\frac{3}{8}$
4. 24.05
5. $40\frac{1}{4}$
6. 2849
7. 1487.5
8. 70.5
9. 31.35

Page 135

	a	b	c
1.	78.5	16.6106	28.26

	a	b		a	b
2.	$254\frac{1}{7}$	616	5.	9856	12474
3.	616	1386	6.	$86\frac{5}{8}$	5544
4.	$38\frac{1}{2}$	$4073\frac{1}{7}$	7.	$6364\frac{2}{7}$	$2\frac{13}{32}$

Page 136

1. 380   4. 78   7. 11304
2. 7000   5. 88   8. 376.8
3. 180   6. 616

184

Page 137

	a	b	c
1.	$22\frac{1}{2}$	99.28	$106\frac{1}{4}$
2.	1728		
3.	$37\frac{1}{2}$		
4.	$17\frac{13}{16}$		

5. 43.2
6. 62.98
7. $65\frac{1}{4}$
8. 198

Page 138

	a	b	c
1.	84	270	68.921
2.	161.28	$55\frac{1}{8}$	180
3.	336		
4.	39.442		
5.	$42\frac{7}{8}$		
6.	14112		
7.	$104\frac{7}{32}$		

Page 139

	a	b	c
1.	1848	396	154
2.	1205.76		
3.	9156.24		
4.	30.85364		

5. 4000.36
6. 3560.76
7. 584.668

Page 140

1. 72
2. 4069.44
3. 3000
4. 3316.625
5. B ; 316.625
6. 7680
7. 100.48

Page 141

	a	b	c
1.	$46\frac{3}{4}$	66	
2.	64	91	54
3.	153.86	119.38	126
4.	343	105	
5.	84.672	1386	

Page 142

1. 6400 ; 5024
2. 1376
3. 251.2
4. 1800000
5. 5400
6. 87.92
7. 3330000

Page 145

1. Trudy ; Sally
2. 16 ; 10
3. Mike ; Marvin
4. E ; D
5. 125 ; 100
6. 625

Photo Credits

Cameron Mitchell, 2, 36, 44

Page 146

1.

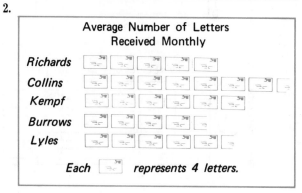

2.

Page 147

1. Carol ; Betty
2. 30 ; 45
3. Alan ; Betty
4. 155
5. 5 ; 1
6. 8 ; 9 ; 18
7. 2, 4, and 5
8. 3 and 6
9. 6

Page 148

1.

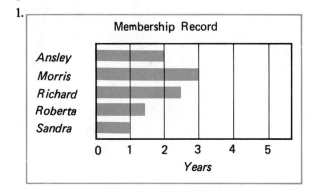

2.

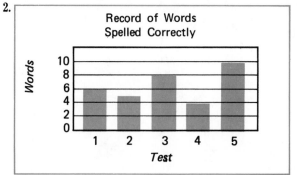

Page 149

1. 70° ; 80° ; 85°
2. 55° ; 9 A.M.
3. 85° ; 1 P.M.
4. increase ; increase ; decrease
5. Aug. ; Dec.
6. Sept. ; Oct.
7. decrease

Page 150

1.

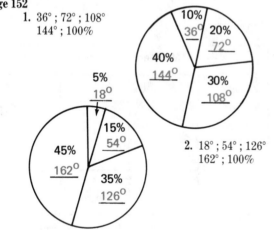

Average Monthly Rainfall (Galveston, Texas)

2.

Bev's Scores on Five Math Tests

Page 151

Have your teacher check your work.

Page 152

1. 36° ; 72° ; 108°
144° ; 100%

10% — 36°
20% — 72°
40% — 144°
30% — 108°

2. 18° ; 54° ; 126°
162° ; 100%

5% — 18°
45% — 162°
15% — 54°
35% — 126°

Page 153

1. 7.50 ; 10 ; 5 ; 2.50
2. 1800 ; 4500 ; 1350 ; 450 ; 900

Page 154

1. 108° ; 72° ;
126° ; 54°

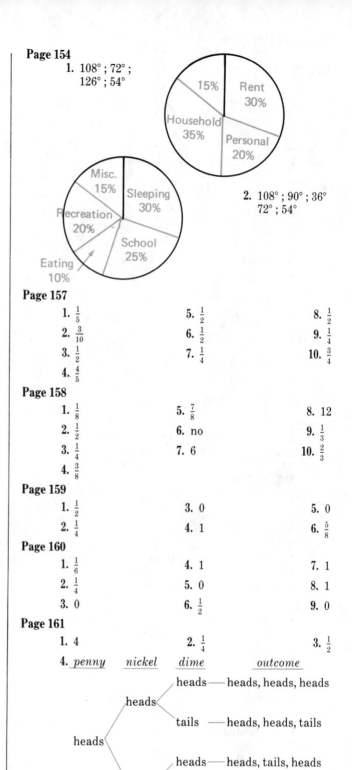

Rent 30%
15%
Household 35%
Personal 20%

Misc. 15%
Sleeping 30%
Recreation 20%
Eating 10%
School 25%

2. 108° ; 90° ; 36°
72° ; 54°

Page 157

1. $\frac{1}{5}$
2. $\frac{3}{10}$
3. $\frac{1}{2}$
4. $\frac{4}{5}$
5. $\frac{1}{2}$
6. $\frac{1}{2}$
7. $\frac{1}{4}$
8. $\frac{1}{2}$
9. $\frac{1}{4}$
10. $\frac{3}{4}$

Page 158

1. $\frac{1}{8}$
2. $\frac{1}{2}$
3. $\frac{1}{4}$
4. $\frac{3}{8}$
5. $\frac{7}{8}$
6. no
7. 6
8. 12
9. $\frac{1}{3}$
10. $\frac{2}{3}$

Page 159

1. $\frac{1}{2}$
2. $\frac{1}{4}$
3. 0
4. 1
5. 0
6. $\frac{5}{8}$

Page 160

1. $\frac{1}{6}$
2. $\frac{1}{4}$
3. 0
4. 1
5. 0
6. $\frac{1}{2}$
7. 1
8. 1
9. 0

Page 161

1. 4
2. $\frac{1}{4}$
3. $\frac{1}{2}$

4.

penny	nickel	dime	outcome
heads	heads	heads	heads, heads, heads
		tails	heads, heads, tails
	tails	heads	heads, tails, heads
		tails	heads, tails, tails
tails	heads	heads	tails, heads, heads
		tails	tails, heads, tails
	tails	heads	tails, tails, heads
		tails	tails, tails, tails

5. 8   6. $\frac{1}{8}$ 7. $\frac{3}{8}$ 8. $\frac{7}{8}$

Page 162

Blue die

	1	2	3	4	5	6
1	1,1	2,1	3,1	4,1	5,1	6,1
2	1,2	2,2	3,2	4,2	5,2	6,2
3	1,3	2,3	3,3	4,3	5,3	6,3
4	1,4	2,4	3,4	4,4	5,4	6,4
5	1,5	2,5	3,5	4,5	5,5	6,5
6	1,6	2,6	3,6	4,6	5,6	6,6

Black die

1. $\frac{1}{36}$
2. $\frac{1}{6}$
3. $\frac{1}{18}$
4. $\frac{1}{12}$
5. 0
6. 1
7. $\frac{1}{6}$
8. $\frac{5}{6}$

Page 163
1. Answers will vary.
2-9. Have your teacher check your work.

Page 164
1. Answers will vary.
2-7. Have your teacher check your work.
8. Answers will vary.
9-10. Have your teacher check your work.

Page 165
1. 20%
2. 100%
3. 25%
4. 75%
5. 50%
6. 25%
7. 0%

Page 166
1. Answers will vary.
2., 4., 6., 8., 10. Have your teacher check your work.
3. 10%
5. 30%
7. 20%
9. 40%
11. 60%
12. 0%
13. 0%

Page 167
1. $\frac{1}{2}$ or 50%
2. 100
3. 500
4. $\frac{1}{4}$ or 25%
5. 5
6. $\frac{3}{4}$ or 75%
7. 45

Page 168
1. 34%
2. 1020
3. 14
4. 420
5. 810
6. 750
7. 1170
8. Verdugo
9. 360

Page 169
1-6. Answers will vary. Have your teacher check your work.

Page 170
1-8. Answers will vary. Have your teacher check your work.

Page 171
1. $\frac{1}{2}$
2. $\frac{1}{3}$
3. 1
4. .25
5. 20
6. 8000

7-8. Answers will vary. Have your teacher check your work.

9. 50%

Page vii
1. 23 ; 3105 ; divide ; 135 ; yes
2. add ; 147.89
3. 23

Page viii
1. add ; subtract ; 376 ; yes
2. 14
3. 15
4. 432
5. 2700
6. 8200

Page ix
1. 28

A-B	A-C	A-D	A-E	A-F	A-G	A-H
B-C	B-D	B-E	B-F	B-G	B-H	C-D
C-E	C-F	C-G	C-H	D-E	D-F	D-G
D-H	E-F	E-G	E-H	F-G	F-H	G-H

2. 15 (p-penny, n-nickel, d-dime, q-quarter)

p	p-n	n-q	p-d-q
n	p-d	d-q	n-d-q
d	p-q	p-n-d	p-n-d-q
q	n-d	p-n-q	

3. 10

| V-W-X | V-W-Y | V-W-Z | V-X-Y | V-X-Z |
| V-Y-Z | W-X-Y | W-X-Z | W-Y-Z | X-Y-Z |

4. Answers will vary.

Page x
1.

3	4	5	6
75	100	125	150

125 ; 8

2.

3	4	5	6
22	26	30	34

30 ; 34 ; third

Page xi
1. 33
2. 5.19 ; 8
3. fourth
4. 22

Page xii
1. 16
2. $7\frac{1}{2}$
3. 180
4. 20 ; 400
5. 90000

Page xiii
1. 14 ; 16
2. 216
3. 9

Page xiv
1. 15
2. 6
3. 15
4. 132000
5. 70 ; 420

6.

Arnie	44	30
Glenn	40	64
Xenia	32	18
Ursula	0	52
Alan	58	10

Page 1

	a	b	c	d	e
1.	7990	17.1	200676	64.05	145.502
2.	5778	3.8	255850	7.37	106.038
3.	$1\frac{1}{10}$	$\frac{5}{12}$	$4\frac{11}{12}$	$2\frac{11}{30}$	$12\frac{5}{24}$
4.	9828	652188	54.864	69.8394	
5.	$\frac{1}{3}$	$\frac{9}{16}$	$1\frac{7}{18}$	$\frac{4}{5}$	
6.	12 r5	11 r103	495	3.25	

Page 2
1. 42850
2. 46735
3. 12775
4. 5919
5. 338

Page 23

	a	b	c	d	e
1.	7821	62.75	69.398	$\frac{1}{2}$	$12\frac{19}{24}$
2.	7527	5.314	5.348	$\frac{1}{12}$	$1\frac{1}{2}$
3.	34371	547056		$\frac{10}{21}$	$1\frac{3}{4}$	$\frac{3}{8}$
4.	17.1432	152.3465	$8\frac{9}{14}$	3	$\frac{7}{16}$
5.	21 r9	3.9	.12	$2\frac{1}{6}$	$1\frac{7}{20}$

Page 24

	a	b	c
1.	$x=6$	$y=5$	$z=16$
2.	$d=15$	$e=42$	$f=52$
3.	$r=5$	$s=22$	$t=8$
4.	$g=12$	$h=17$	$j=30$
5.	$m=18$	$n=7$	$p=45$
6.	$a=12$	$b=11$	$c=17$
7.	$u=42$	$v=0$	$w=16$

Page 37

	a	b	c
1.	$m=10$	$n=15$	$p=11$
2.	$r=42$	$s=195$	$t=86$
3.	$a=27$	$b=21$	$c=30$
4.	$x=36$	$y=50$	$z=48$
5.	$w=40$	$x=540$	$y=67$
6.	$d=12$	$e=51$	$m=6$

7. $\frac{1}{4}x=5$; 20

Page 38

	a	b	c
1.	$9x$	$10y$	$3z$
2.	$4a$	$4b$	$3c$
3.	$r=9$	$s=7$	$t=13$
4.	$d=20$	$e=34$	$f=50$
5.	$u=3$	$v=5$	$w=8$
6.	12	7. 216	

Page 47
1. $x+3x=32$; 8
2. $x+(x+9)=87$; 48
3. $x+(2x+5)=215$; 70 ; 145
4. 24
5. 153
6. 3

Page 48

	a	b		a	b
1.		$\frac{7}{8}=\frac{28}{32}$	3.	$\frac{7}{9}=\frac{21}{27}$	$\frac{24}{15}=\frac{8}{5}$
2.		$\frac{2}{3}=\frac{10}{15}$	4.	$n=1$	$n=25$
5.	$n=30$	$n=4$	9.	1	60
6.	$n=5$	$n=6$	10.	80	60
7.	$n=20$	$n=56$	11.	4.5	75
8.	4.32	$43\frac{3}{4}$	12.	5.561	150

Page 67

	a	b			
2.	5 to 4	$\frac{5}{4}$	4.	5 to 9	$\frac{5}{9}$
3.	4 to 30	$\frac{4}{30}$	5.	$n=1$	$n=5$

Page 67 (continued)

6.	$n=40$	$n=56$

	a	b
7.	$n=12$	$n=3$
8.	22.4	20

9.	72	400
10.	70	$3\frac{3}{8}$
11.	270	25

Page 68

	p	r	t	i
1.				$22.40
2.				$8.25
3.			1 year	
4.		8%		
5.	$600			
6.		9%		

	total amount
7.	$224.72
8.	$115.76
9.	$337.46
10.	$420.38

Page 77

	p	r	t	i
1.				$67.50
2.				$119
3.		12%		
4.			$2\frac{1}{2}$ years	
5.	$3840			
6.		15%		

	total amount
7.	$343.47
8.	$694.58
9.	$562.75
10.	$409.07

Page 78

1.	liter
2.	centimeter
3.	milligram
4.	.001

5.	1000
6.	.01
7.	5
8.	65

	a	b
9.	.001	10
10.	2000	2000
11.	500	3
12.	1400	50

Page 87

	a	b
1.	5	
2.	35	
3.	250	350
4.	6000	.007
5.	2.6	.6
6.	7500	.25
7.	12000	.0135
8.	5400	1.2

	a	b
9.	45	.26
10.	58000	.4
11.	3	3800
12.	600	.05
13.	0 ; 32	
14.	100 ; 212	
15.	2300 ; 2.3	
16.	Inez ; 7	

Page 88

	a	b
1.	4	56
2.	2	16
3.	180	83

	a	b
4.	2	304
5.	4	14
6.	8	15

	a	b	c
7.	10 ft 7 in.	2 min 22 sec	8 lb 12 oz
8.	8 yd 1 ft	2 gal 3 qt	8 lb 12 oz
9.	8320	8300	8000
10.	74490	74500	74000

Page 103

	a	b
1.	36	11
2.	$1\frac{1}{2}$	75
3.	80	147

	a	b
4.	4	45
5.	12	7

Page 103 (continued)

	a	b	c
6.	5 lb 14 oz	6 ft 2 in.	6 min 42 sec
7.	1 gal 3 qt	14 yd 2 ft	9 hours 38 min
8.	4770	4800	5000
9.	63580	63600	64000
10.	11645 ; 12000	5068 ; 5000	23088 ; 24000

Page 104

	a	b	c
1.	l	k	a, h
2.	b	i	f
3.	d	h	j

	a	b	c
4.	g	c	a
5.	e	a	f
6.	60	90	

Page 113

	a	b	c
1.	$\overline{QP}$ or $\overline{QR}$	$\overline{PR}$ or $\overline{RP}$	
2.	$\triangle ABC$	$\triangle DEF$	
3.	$\triangle ABC$ and $\triangle JKL$	$\triangle JKL$	
4.	90, right	40, acute	105, obtuse
5.	parallel	perpendicular	perpendicular

Page 114

1.	12
2.	7
3.	10

	a	b
4.	10	8
5.	10.63	11.31

Page 129

1.	12	4. 2	7. 11.36	10. 9.9
2.	8	5. 10	8. 12.04	
3.	12	6. 63	9. 6	

Page 130

	a	b	c
1.	24	24	33
	35	22.5	$86\frac{5}{8}$
2.	36	28	27.2
	52	40	46.24

	a	b
3.	125	142.968
4.	452.16	120

Page 143

	a	b	c
1.	24	32	28
	36	$45\frac{1}{2}$	42.24
2.	30	62	63
	30	$157\frac{1}{2}$	222
3.	4 ; $25\frac{1}{7}$; $50\frac{2}{7}$		
4.	10 ; $31\frac{3}{7}$; $78\frac{4}{7}$		

	a	b
5.	1384.74	2160

Page 144

1.	16 ; 12	5.	15 ; 10 ; 20
2.	Jan. ; Apr.	6.	2 ; 5
3.	30	7.	Feb. ; Jan.
4.	3 ; 2	8.	decrease

190

Page 155

1. Rosie   2. Jay   3. 6 ; 12 ; 9

4.

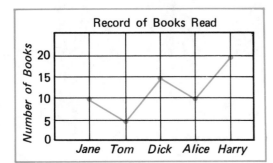

Record of Books Read

5. 144° ; 18° ; 36° ;
   54° ; 108°

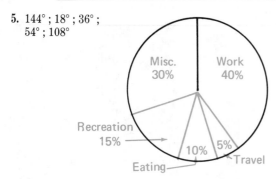

Page 156

1. $\frac{1}{3}$ 3. $\frac{1}{2}$ 5. 0

2. $\frac{1}{6}$ 4. 1

6.

penny	nickel	outcome
heads	heads	heads, heads
	tails	heads, tails
tails	heads	tails, heads
	tails	tails, tails

7. 80%   9. 75%   11. 150

8. 2500   10. 15

Page 172

1. $\frac{1}{6}$ 5. $\frac{2}{3}$ 8. 9000

2. 0   6. 50%   9. 40%

3. $\frac{1}{3}$ 7. 75   10. 8

4. $\frac{1}{2}$

11.

penny	dime	outcome
heads	heads	heads, heads
	tails	heads, tails
tails	heads	tails, heads
	tails	tails, tails

Page 173

	a	b	c	d	e
1.	93748	64614	86.604	.1049	830.61
2.	487506	81.38	105	1257.5	.402976
3.	$2\frac{1}{8}$	$\frac{5}{12}$	$2\frac{7}{24}$	$17\frac{1}{2}$	$\frac{3}{4}$

	a	b	c
4.	$b=6$	$a=104$	$d=91$
5.	$h=26$	$j=8$	$s=360$
6.	$a=10$	$n=6$	$r=7$

Page 174

	a	b	c
7.	$n=6$	$n=36$	$n=9$
8.	.25 ; 25%	$\frac{1}{10}$; .1	
9.	6	48	50

10. 111   11. 240

12. $x+x+12=150$; 81   12. $80\times6=10\times W$; 48

Page 175

	a	b	c	d	e
1.	20951	94	107.413	45.93	.01014
2.	26784	207	24	5	$1\frac{5}{6}$
3.	$1\frac{1}{3}$	$\frac{2}{3}$	9 yd 1 ft	1 pt 1 cup	2 min 45 sec

	a	b	c
4.	$n=4$	$a=129$	$e=3$
5.	$h=19$	$j=8$	$s=96$
6.	$a=10$	$n=10$	$r=2$
7.	$n=4$	$n=63$	$n=15$

Page 176

	a	b	c
8.	.5 ; 50%	$\frac{3}{4}$; .75	
9.	20	140	24
10.	600	8000	.1
11.	.1	2300	5.6
12.	5	6	255
13.	4000	2	14
14.	30000	32500	32550
15.	3000 ; 3265	30000 ; 31070	27000 ; 28613
16.	0	32	

17. d   18. f   19. e   20. a

Page 177

21. 6   22. 15

	a	b	c
23.	$37\frac{1}{2}$	200.96	68.87
24.	72	91.125	7700

25. April; August   26. April, May, June

Page 178

27. $\frac{1}{6}$; $\frac{5}{6}$ 31. 4

28. 5 ; 25   32. $x+x+2=40$; 21

29. 800   33. $80\times6=8\times W$; 60

30. 600